Leadership and Transformation of the Army Chaplaincy During WWII:

DOTMLPF-P as an Assessment Tool For Religious Support

Office of the Chief of Chaplains
Department of the Army
Washington, D.C. 20310

July 15, 2020

ISBN 978-0-578-86298-9
90000>
9 780578 862989

ISBN: 978-0-578-86298-9

Second Printing 2021

Cover images:

Chaplain Paul Maurer performs the burial service for Major General Maurice Rose, April 2, 1945. (http://www.bensavelkoul.nl/Generaal_Maurice_Rose.htm, accessed April 19, 2019)

Cover design: Jeff Gerke

Acknowledgments

I am thankful for the following people who assisted me with this project. Colonel Frank Frazier, who guided me in my initial Army War College paper. Chaplain (Brigadier General) William Green Jr., Chaplain (Colonel) David W. Lile, and Chaplain (Colonel) Steve E. Cantrell who provided technical feedback for my initial paper. Their critique was sound and pastoral. I am truly grateful for their candor and professionalism. I am also thankful for Nancy Waldron, Chaplain (Colonel) Eddie Kinley, and Lieutenant Colonel (Retired) Edwin Kennedy who provided input and assisted with the transcript of this book, and to the Army Chaplain Historian Mr. Daniel Fitzpatrick for his editorial comments and assistance with the publication of this manuscript.

I am grateful for my parents who nurtured my interest in history. Thanks to my Mom, who bought my first book, *Meet Abraham Lincoln* by Barbara Cary. Thanks to my Dad, who took me to battlefields and let me roam the halls of the Smithsonian National Museum of American History. Thank you to my wife and children for tolerating the "war room" of regulations, manuals, papers, and artifacts. I love you.

The author dedicates this work to the World War II Chaplains and Chaplain Assistants, some I met personally, and many I feel as if I know them through their writings. I am thankful for their service, inspiration, and sacrifice so that my family has the freedom to worship freely and without fear. To them, I am grateful.

List of Figures

Table of Contents

Foreword

Since the founding of our nation, the United States Army Chaplaincy has provided for the free exercise of religion for our soldiers. The weapons and methods of warfare have changed and will continue to change, but the mission of Army chaplains is unchanging—ever since General George Washington insisted in 1775 that the Continental Congress provide one ordained chaplain in each regular regiment of the Continental Army, ensuring those American Soldiers would always have spiritual support and guidance near at hand. While nurturing the living, caring for the wounded, and honoring the fallen, the Chaplain Corps continues to care for the soul of our Army today, just as our chaplains did in 1775.

The Army Chaplaincy consists of men and women of unique calling, identity, and leadership responsibilities. No one else can do what the members of the Chaplain Corps do for the Army. We build Army readiness by delivering direct spiritual and religious Soldier and Family care, and by providing moral leadership across the Army's formations at every echelon, everywhere and anywhere America's Solders are required to perform their duties—while our nation is at peace and while we are at war, at home and abroad, in garrison and on the battlefield.

This book captures the history of the World War II Army Chaplaincy in a way that will inspire agile and adaptive religious support leaders to continue to advance the Chaplain Corps' proud legacy of sacred service. Our identity as Chaplain Corps members leads us to principled, meticulous actions of care for Soldiers, their Families, and Army civilians. We do not perform this function alone, but as an integral and critical part of the larger team responsible for enhancing the readiness of Soldiers of all ranks. We do it for God and Country!

Chaplain (Major General) Thomas L. Solhjem
U.S. Army Chief of Chaplains

Introduction

In the 1942 book *Prelude To Victory,* the Scottish-born, American journalist and author James B. Reston quoted General Douglas MacArthur's philosophy of warfare as he saw it in 1940.

> The history of failure in war can almost be summed up in two words: 'Too Late.' Too late in comprehending the deadly purpose of a potential enemy; too late in realizing the mortal danger; too late in preparedness; *too late in uniting all possible forces for resistance; too late in standing with one's friends* [*sic*]. Victory in war results from no mysterious alchemy or wizardry but depends entirely upon the concentration of superior force at the critical points of combat.[1]

The United States was not at war when MacArthur wrote this. Yet, he understood and would soon realize the impact of not being prepared for war. In MacArthur's words, he uses "alchemy or wizardry" to explain the difficult task of combining human effort within civilian and military systems in order to achieve a successful outcome. The "mysterious alchemy or wizardry" is the art and science of war; of linking the strategic, operational, and tactical levels of war; to visualize, describe, and direct military operations; to plan, prepare, and execute while continually assessing all military operations. The outcome of these meticulous and demanding systems results in victory or defeat; life or death; or the survival and destruction of a nation. In the history of warfare, military leaders have struggled with utilizing military systems with the requirement to be ready to fight tonight to defend their nation.

MacArthur's alchemy or wizardry is analogous to the Joint Capabilities Integration and Development System (JCIDS) process. JCIDS is the art, science, and transformational lens for the United States military acquisition process. Within JCIDS are the domains of Doctrine, Organization, Training, Materiel, Leadership and Education, Personnel, Facilities, and Policy (DOTMLPF-P). These domains provide the capabilities assessment for the Army's acquisition process.

The purpose of this work is to use Doctrine, Organization, Training, Materiel, Leadership and Education, Personnel, Facilities, and Policy (DOTMLPF-P) to assess the transformation and effectiveness of the Army Chaplaincy led by Chaplain (Major General) William Arnold during World War II. This transformation contributed to effective religious support to all soldiers during World War II. This work will show the benefit of the domains for transformation and readiness. The context is the United States Army Chaplaincy from 1937-1945. Examples within each domain will show how the Army Chaplaincy transformed during World War II in order to provide religious support.

Current military leadership is asking questions from historians regarding principles and procedures for large-scale, prolonged conflict. Capturing these events and examples in the modern framework of DOTMLPF-P will show the power and relevancy of the entire JCIDS process. These examples are not mere trivial chaplain anecdotes, but legitimate ways and means to provide religious support by caring for service members and their families. Former Chief of Chaplains, Chaplain (Major General) Gaylord T. Gunhus, best summarizes the essence and methodology of this work.

> In some sense, history is also prophecy; for it portrays issues which have recurred periodically in the past and which may well appear again in the future. The solutions of the past, especially within defined, traditional institutions, may well suggest the parameters of possibility for future choices. History, therefore, is not a field to be scanned egocentrically, but to be studied systemically and ecumenically for its practical lessons.[2]

The intent is not to replicate the actions of the World War II chaplaincy, but to inspire creative thinking based upon successful historical solutions.

The historiography of the Army chaplaincy falls into three categories: hagiography, social-psychology, and scholarly.[3] This book is not the typical chaplain history book, nor does it f t easily within any of the three genres. As with many current doctrinal publications, this work will present historical actions within a current doctrinal framework. This book recognizes the strengths and avoids the weaknesses of previous chaplain history genres, while including current doctrinal references.

Before the Vietnam War, many works by or about chaplains were hagiographic in nature. Chaplains could do no wrong while serving their flock wherever they may be. Many chaplains did perform incredible deeds on the battlefield. However, many books written by or about chaplains avoided the harsh realities of war. These works more closely resembled a Michelin travel map through Europe than a historical account of what happened.

The second category began shortly after the Vietnam War and focused upon the social psychology of the Army chaplain. Post-modernist writers and chaplain autobiographies emphasized the impact of war and post-traumatic stress. These books present chaplains as conflicted souls who are tormented by the dichotomy of serving a loving God while working in an enterprise whose mission also includes death and destruction. In this genre, the chaplaincy is a secular institution serving the needs of the State instead of the higher calling of the profession. "Presentism," the introduction of current values to evaluate and analyze historical actions is prevalent in this genre. The lack of historical context and selectively chosen data fills the social-psychology history of the Army chaplaincy. However, the value of the social-psychology history of the chaplaincy is the critical nature of the writings. Blind spots and vulnerabilities are in every organization. Reading this genre provides insight for potential requirements for the gap analysis for the JCIDS process.

The third and smallest category is scholarly. These works do not focus just upon the miraculous on the battlefield or select few at the higher echelons of the Army. Nor are they void of any context. These works analyze the data within the historical context. Missing is any form of religious bias in the interpretation of the historical event. Authors in this genre highlight success and failures at every level within the chaplaincy. This genre does not include the cynical view of chaplains but provides proven methods to assist the resilient chaplain. Conclusions are relevant and sound for current application to allow for religious freedom. Absent from this genre is current military doctrine that could assist with placing historical actions in current context.

Dr. John Mark Mattox at the 2018 International Military Ethics Symposium said, "A hundred years from now, no one except military history buffs and a few academics will care about your tactics. What everyone will care about is whether what you did was right."[4] This work will provide examples of how the Army Chaplaincy provided religious support and did what was right. The content of this work does not second-guess the leadership in the field. It does not apply 21st Century morals and standards upon the World War II generation. It does, however, lay out the morals and standards for that generation and explain how the chaplaincy fulfilled its sacred calling to provide for the religious needs of a nation at war.

This work is unique within chaplain historiography. It applies a systems approach utilizing the modern military methodology of JCIDS to analyze and assess what military leaders wrote and acted upon in order to provide religious support for personnel in the U.S. Army. Instead of battle maps showing troop locations and their movement, this work will use quotations from chaplains and Army chaplaincy publications to show, in the words of the writers themselves, how they articulated religious support operations. Since chaplains train and teach human interaction skills, it is imperative to see how chaplains interacted with one another to provide religious support.

A systems approach and the utilization of the JCIDS process provides a framework for assessments of past or current military operations, and as a tool in the development of future religious support plans and pilot programs. A systems approach provides a "common operating picture" that helps leaders "see ourselves" and therefore drive effects. However, there is a tendency with a systems approach where action officers and working groups view the process, charts, and diagrams as success. The process or the charts must not enamor leaders. Nor should the systems and charts manipulate leaders. Successful leaders must understand and master the systems that keep people accountable, and create unity of effort toward the endstate.

In this work, the first chapter will provide an overview of the JCIDS process and the importance of using DOTMLPF-P as an assessment tool for readiness and transformation. Then a survey of how the Army chaplaincy used DOTMLPF-P in the past. The first chapter will conclude by providing the context of World War II and a short biography of the Chief of Chaplains, Chaplain (Major General) William Arnold, who led the chaplaincy during World War II. The following chapters will begin with the Army definition for each domain and then offer a suggested religious support definition for each domain. Each chapter will evaluate the domains within DOTMLPF-P respectfully and show the "capability gap" of what the Army chaplaincy started with, and how it closed the gap within each domain. The examples of the capability from each domain within the DOTMLPF-P framework will show the sequence of actions (Ways) and the resources used (Means) toward the objective (Ends) of how the chaplaincy provided religious support during World War II. The last chapter will include a brief synopsis of Chaplain Arnold's legacy and the importance of using DOTMLPF-P during the lean years between World War II and the Korean War. Some chapters are more robust because of the complexity of the domain, while other chapters, such as the chapter that examines facilities within the DOTMLPF-P framework only have the Army chapel as an example, despite the critical requirement for chapels when providing religious support.

Included in this work are two examples of Force Modernization and Transformation. Appendix 1 is the Army Force Modernization Proponent System with suggested Religious Support definition for each domain, linked to Religious Support capabilities, competencies, and functions. This logical flow will ensure the consideration of all domains and their role in the performance of religious support. Design Methodology contributes to identifying requirements. In Appendix 2, "Design Approach to Chaplain History (1937-1945)" is a way the Office of the Chief of Chaplains (OCCH) during World War II could have used Design in order to develop an approach to the complex issues before World War II.

In summary, this is a systems approach to capture and measure Chaplain Corps' core competencies of nurturing the living, caring for the wounded, and honoring the fallen;[5] while performing the capabilities of advising the Command and providing religious support for all soldiers and families wherever they may be.[6] A study of the transformation of the Army Chaplaincy during World War II will show how Chaplain Arnold and thousands of other chaplains epitomized the freedom to provide for the free exercise of religion for all soldiers.

Notes

[1] James B. Reston, *Prelude To Victory,* (Alfred A. Knopf New York 1942), 64.

[2] John W. Brinsfield, Jr., *Encouraging Faith, Supporting Soldiers: The United States Army Chaplaincy, 1975-1995*, (Washington, DC: Office of Chief of Chaplains, 1997), 283.

[3] The author created the three categories for chaplain historiography. They represent similar classifications for other genres in academic writing. There could be a fourth category of "historical fiction." These works are autobiographical, biographical, or topical in nature. Fiction replaces facts. Pseudonyms, ad hominem and self-interest are in the historical fiction genre.

[4] John Mark Mattox, Ph.D, "International Military Ethics Symposium" (National Defense University, Wash. D.C., August 1, 2018).

[5] U.S. Department of the Army, Army Regulation 165-1, *Army Chaplain Corps Activities*, (Department of the Army, June 23, 2015). "Honor the dead" was the term used since 1984. "Honor the fallen" is the current Army Regulatory term.

[6] U.S. Department of the Army, Field Manual 1-05, *Religious Support,* (Department of the Army, January 2019) 1-4.

Chapter 1

DOTMLPF-P As An Assessment Tool For Religious Support During WWII

The age-old question of how to do something better and more quickly resonates with every society. For the military, this question could mean life or death for its soldiers. Transformation and readiness are the overarching concepts of how the military does things quicker, better, and with limited resources.

Today's United States military uses the Joint Capabilities Integration and Development System (JCIDS) to answer transformation and readiness questions within the acquisition process. Applying JCIDS as an assessment framework to a historical event provides insight for current and future military operations.

Elliott Converse's classic work, *Rearming for the Cold War, 1945-1960, History of Acquisition in the Department of Defense*, describes how World War II was a "watershed" for the United States. "The war was the catalyst for far-reaching economic and social changes, including the industrialization of the South and the West and equality for women and [African Americans, ed.]."[1] The Army chaplaincy with Chaplain (Major General) William Arnold as the Chief of Chaplains was not a mere witness, but a key participant and a leader of change. Doctrine, Organization, Training, Materiel, Leadership and Education, Personnel, Facilities, and Policy (DOTMLPF-P) will help understand these changes within the Army chaplaincy during World War II and demonstrate their relevancy today.

Joint Capabilities Integration and Development System

Joint Capabilities Integration and Development System (JCIDS) is a Department of Defense (DOD) process that provides the Joint Force with the capabilities needed to perform across the full range of military operations of offense, defense, stability operations, and defense support of civil authorities. JCIDS, along with the Defense Acquisition System (event-driven) and the Planning, Programming, Budgeting, and Execution (calendar-driven) process form the principal DOD decision support processes for developing and acquiring capabilities required by the military forces to support the national defense strategy. The domains of Doctrine, Organization, Training, Materiel, Leadership and Education, Personnel, Facilities, and Policy (DOTMLPF-P) are the framework within JCIDS to assist leaders with the acquisition process.

In the JCIDS process, Combatant Commanders receive strategic guidance and assess current warfighting capabilities in order to confront current and future threats to United States national security. These threats include state and non-state actors, and environmental emergencies. Combatant Commanders identifies "capability gaps" within the domains of DOTMLPF-P. Non-materiel and materiel solutions fill the "capability gaps" in order to defeat threats against the United States. Cheaper non-materiel solutions should be the first course of action to fill the "capability gap." Materiel and facility solutions fill the "capability gap" when no other non-materiel items within the force are available.

JCIDS is also an accountability tool that shows Congress how the military is conducting due diligence in the budgetary and acquisition processes, in order to defeat current and future threats. However, the benefits of JCIDS and DOTMLPF-P go much further than national security and the halls of Congress. The most important benefit of utilizing DOTMLPF-P is when, as a framework, it creates unity of effort by the linking Command guidance with the strategic and operational environment in order to achieve specified short-and long-term objectives.

Competent and experienced leaders visualize requirements within DOTMLPF-P. To assist with the art of achieving objectives, leaders use several other frameworks as well. PMESII-PT (political, military, economic, social, information, infrastructure, physical environment, and time) describe conditions of an operational environment in terms of operational variables. METT-TC (mission, enemy, terrain and weather, troops and support available, time available, and civil considerations) takes the operational variables into mission variables during mission analysis. The CARVER analysis (Criticality, Accessibility, Recuperability, Vulnerability, Effect, and Recognizability) prioritizes enemy targets and assesses friendly targets for protection. These additional frameworks assist leaders in developing requirements for each domain within DOTMLPF-P. Leaders must validate each requirement within each domain to ensure it will meet current and future readiness demands.

Readiness and Transformation

Readiness is essential to every military.[2] Maintaining the edge over the adversary requires anticipating and transforming for current and future requirements. The Research and Development Corporation (RAND) sums this up under "military transformation—the adjustment of a nation's military to achieve a specified objective—analyzes four major issues: force structure, modernization, readiness, and sustainability."[3] Force structure managers, along with modernization and sustainment integrators, use DOTMLPF-P as a framework in their analysis for readiness. However, defining and assessing readiness is problematic.

Readiness assessment is subjective and is usually self-assessed and reported semi-annually as either "ready" or "not ready." Readiness as a term is an abstraction. The only scientific measurement for assessing readiness is medical readiness. However, just because a soldier is medically ready for deployment does not mean the soldier will perform in combat. Without identifying what it is the leader is trying to achieve to be ready, there is no context from which to evaluate whether a soldier is ready or not. Soldiers must have the skills trained in specific tasks in specified conditions. Soldiers must have the tools or equipment to perform the tasks. Soldiers must have the leadership and the motivation to perform the tasks. Soldiers must have the capacity and the sustainability to perform the tasks over a long period. A soldier that is ready, trained, skilled, equipped, willing, and capable to fight would be of minimal value in the Army if the doctrine and organization cannot support and sustain the soldier.[4] Therefore, the soldier's condition and personal readiness is the linchpin for the successful application of other domains within DOTMLPF-P and essential to achieving objectives at every echelon of war. The often-overlooked personal self-development training domain is paramount to mission readiness and applies to every soldier at every level of war. Therefore, personal readiness is essential to mission readiness.

Mission readiness, and in particular personal readiness, is nothing new and is essential for the survival of any nation. Soldiers on alert at Fort Bragg or the soldiers who are ready to "fight tonight"[5] on the Korean Peninsula are essential for mission and personal readiness. The importance of personal readiness was evident in the lessons from World War I. Colonel James A Moss's *Manual of Military Training* was the standard for training soldiers for World War I. The first edition in 1914 contained no prelude. In his second edition in 1917, Moss incorporated lessons from the British and added a prelude that begins by stating that the objective of all military training is to win battles. Moss then describes the importance of training tasks, equipment, discipline, leadership, and finally a diagram and description of the soldier.[6] For Moss, the personal wellbeing of the soldier is the key to winning battles. It is no coincidence that at Fort Dix, where soldiers mobilized for both World Wars, is a statue of the Army soldier titled, "The Ultimate Weapon." Training, equipping, and protecting the soldier has been and will always be the key factor for every successful military.

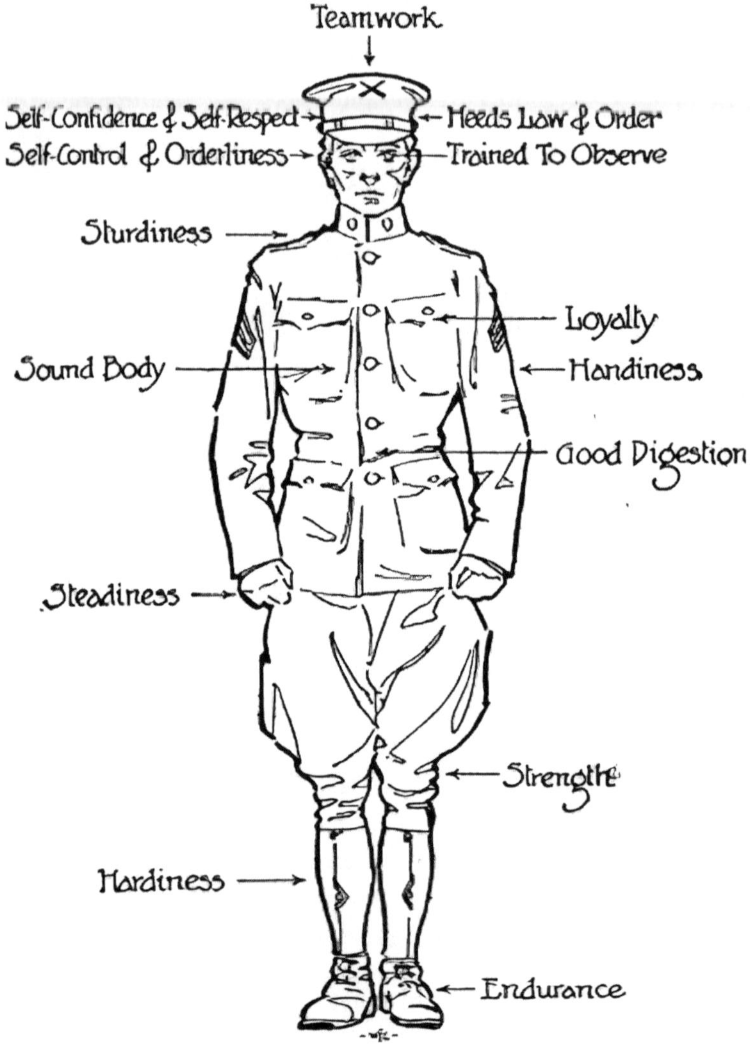

Figure 1. Illustration from Colonel James A Moss's Manual of Military Training
Source: James A. Moss. Manual of Military Training, Second Revised Edition.

Before the discovery of Penicillin, a soldier's inclination to getting venereal disease meant that the infected soldier became temporarily or permanently ineffective for combat. Therefore, the Army's number one priority toward the end of World War II was the education and prevention of venereal diseases.[7] Drug use among soldiers during the Vietnam War created significant readiness, moral, and morale issues. In all of these situations, Chaplains in the Army provided leadership and guidance for the "ultimate weapon" to protect and prepare them in their readiness for military operations. Chaplain leadership and instruction included *The Chaplains Hour* publication, character guidance lectures, moral heritage programs, family readiness presentations, and pre-deployment and post-deployment briefings. Army chaplains were able to do this since they are trained and certified to advise and care for the soldier's wellbeing. Also, Army chaplains and their assignment at battalions are the links to behavior health specialists who are assigned at the brigade combat team and higher echelons. Finally, chaplains provide the dual function as both a secular and spiritual counselor. Therefore, the Army chaplain fulfills four of the five dimensions of Personal Readiness, which are Physical, Psychological, Social, Spiritual, and Family Preparedness.[8] Utilizing DOTMLPF-P to evaluate how chaplains provide support in the future will enhance soldier readiness throughout the all-volunteer Army.

Today, the all-volunteer force is vastly different from the WWII draft Army. However, every chaplain in the history of the chaplaincy, including WWII, answered the call and volunteered for service. Every Army chaplain today stands in the proud tradition of thousands of other Army chaplains who answered the call to serve God and Country. When chaplains serve their country, they represent the best the country has to offer by supporting freedom and the free exercise of religion and in some cases, a soldier's right to not worship. The chaplains are there to advise in their responsibility to support the free exercise of religion, and to support the soldier and their family regardless of their belief system. DOTMLPF-P as an assessment tool will assist chaplains at every echelon to care for soldiers and their families.

DOTMLPF-P provides insights regarding the transformation and readiness of the Chaplain's Corps during World War II. The examples provided are not exhaustive for each domain. Further, most of the examples listed in each domain may overlap with other domain capabilities. For example, the Policy of Privileged Communication, which protects the communication between a chaplain and a service member, could also be a capability under the Doctrine domain since privileged communication is part of providing religious support. Army Regulation 5-22, *The Army Force Modernization Proponent System* and JCIDS define each domain within DOTMLPF-P. This work will provide suggested religious support definitions for each domain in order to determine capability gaps for religious support.[9]

Leaders during WWII did not have DOTMLPF-P as a framework to solve "capability gaps." Therefore, the use of DOTMLPF-P applied to the World War II Chaplain Corps will provide insights into the character of these chaplains and lessons for current and future Army Chaplains.

Army Chaplain Attempts and Opportunities at DOTMLPF-P

In the past, the chaplaincy produced several publications revealing how the chaplaincy utilized the DOTMLPF-P system. *The Army Chaplaincy*, a professional bulletin of the Unit Ministry Team, Summer-Fall 1999, provided an overview of Doctrine, Training, Leadership, Organization, Materiel, Soldier (DTLOMS), an earlier version of the DOTMLPF-P process.[10] This document presents the importance of nesting religious support within the strategic guidance and military organizations, but it also reflects the tendency in this process to address the means, and not the ways to address the gaps analysis. At the turn of the 21st century, the draft yet-unpublished *TRADOC* Pamphlet 525-1-05 *Transforming the US Army Chaplains for the Modular Army* did an excellent job of identifying gaps and ways to decrease the gaps.[11] However, the ways and terminology were not within Army doctrine and accepted terminology. In January 2012, OCCH began the "Establishment of Policy Concerning Structure and Functioning of Chaplaincy Participation in Process Integration Concept (PIC).[12] The PIC concept used the Resource-informed, Integration-focused, Outcome-based model framed within DOTMLPF domains for solution sets. From January 2014 to the fall of 2016, Installation and Management Command (IMCOM), Senior Chaplain Assistant [Chief Religious Affairs Non-Commissioned Officer, ed.], Sergeant Major Pamela Wilson started the "Fifty-Six Mike" monthly newsletter.[13] The title, "Fifty-Six Mike" (56M) is the Military Occupational Specialty (MOS) for the Religious Affairs Specialists. The IMCOM Command Chaplain contributed to each newsletter, expressing the importance of the IMCOM mission and his appreciation for the team approach to religious support. Sergeant Major Wilson used DOTMLPF as an outline to educate, provide information, and solicit feedback. This newsletter included doctrinal references, advice, and good news stories that included pictures of chaplain assistants.

Recently, in a few cases, leaders have used DOTMLPF-P as a framework to conduct an After Action Report (AAR). For large operations, such as Combat Training Center rotations or conducting Vacation Bible School, DOTMLPF-P does provide a comprehensive tool to assess and improve future operations. However, for small tactical operations like a raid or a worship service, DOTMLPF-P is cumbersome and will "inhibit Soldiers who seem reluctant to participate" in the AAR process.[14]

A modified Capability Mission Lattice found in the Defense Acquisition University's *Manual for the Operation of the Joint Capabilities Integration and Development System* (JCIDS) will assist with religious support readiness and transformation (See Appendix 1). The strength of this modified Capability Mission Lattice is the logical flow from the strategic guidance toward the strategic endstate. Army Regulation 5-22, *Army Force Modernization Proponent System*, provides definitions for each domain. Linked to these definitions are suggested religious support definitions for each domain. The Capability Mission Lattice utilized the Universal Joint Task List (UJTL). The UJTL is a comprehensive list of tasks and does provide a variety of scenarios and contexts. However, a drawback of the UJTL is the binary determination that the organization that is either "ready" or "not ready" when it performs the specified task. Missing is an analysis of the valuable contribution to readiness by the creative and agile leader exercising mission command philosophy and disciplined initiative. Therefore, to correct this issue in the Capability Mission Lattice for Religious Support, the UJTL now has the religious support capabilities, competencies, and functions linked with strategic guidance toward known and unknown threats. This logical flow will ensure the consideration of all domains in order to execute religious support. Religious support leaders who use this Capability Mission Lattice will exercise mission command when they understand, visualize, describe, direct, and execute religious support.

Including DOTMLPF-P into other strategic tools such as the center of gravity analysis and Ends, Ways, Means, Risk, will contribute to a feasible, acceptable, suitable, distinguishable, and complete outcome. Center of gravity analysis will assist in identifying capabilities and requirements for religious support. For example, what are the capabilities, requirements, and vulnerabilities to conduct religious support in the garrison, or an operation? Approaching this problem set utilizing the DOTMLPF-P domains will assist in the gap analysis. Ends, Ways, Means, Risk, are essential to formulating a strategy. DOTMLPF-P helps the strategist develop and prioritize objectives based upon short-term events; and Planning, Programming, Budgeting and Execution calendar driven capabilities, and requirements. DOTMLPF-P provides essential details for effective and enduring strategies for religious support.

The Context, 1937-1945

World War II stood in stark contrast to World War I. World War I, for the most part, was a family feud among nations who had similar religious beliefs. Royal families who were vice regents for the church were still settling the score and setting boundaries from the Peace of Westphalia. France is the notable exception, as a Royal did not rule it, but it suffered most from German unification in 1871. However, France still presented its conflicts in religious terms. The family feud of World War I resulted in the rise of a new religion of totalitarian secularism, void of freedom and human rights. Communism in the Soviet Union, Fascism in Italy, National Socialism in Germany, and the rise of the Showa era in Japan represented totalitarian regimes that limited freedom, particularly, religious freedom.

In the 1930s, the United States continued to wrestle with biases and prejudices amongst the population. Post-reconstruction South forced segregation against African Americans. The economic depression of the 1930s saw the rise of hate organizations throughout the United States. Suspicion and wounds from the Indian Wars created mistrust between government authorities and the native American population. The rise of totalitarian secularism void of freedom and human rights forced the United States to face its hypocrisy. Does the United States honestly believe and practice freedom, liberty, and justice for all as the Founding Fathers wrote about? The United States answered this question by promoting the principles of the Founding Fathers and implemented changes during World War II. The Army Chaplaincy was a part of this change.

The free exercise of religion is part of the foundation of the United States. Defending freedom was at the forefront of the minds of the founding fathers in 1776, as they gathered in Philadelphia, Pennsylvania. The history of religious freedom and Philadelphia are no strangers. Philadelphia was the second largest English-speaking city in the world and the second largest British port when the Second Continental Congress met in 1775. Earlier, in 1681, William Penn created the city of Philadelphia. Penn, a Quaker, received a large portion of land from his father, Admiral Sir William Penn. Persecution of Quakers in Europe spurred Penn to allow all faiths to gather "in brotherly love," hence the name Philadelphia. In this area, in addition to Quakers, there was one of the earliest Roman Catholic Churches, which is still in service today and stood in contrast to the Pennsylvania Dutch, borne from the radical Reformation in Germany. Philadelphia and the surrounding area became the beacon for a diverse community that included the Quakers, Pennsylvania Dutch, Roman Catholics, Episcopalians, Presbyterians, Methodists, Baptists, and Congregationalists.[15] The fertile soil of liberty and religious freedom allowed for the melting pot of civil citizenry and robust economic prosperity.

During 1776, in the city of Philadelphia, the Second Continental Congress created the Board of War, also known as the Board of War and Ordnance, and later as the War Department.[16] The seal for this organization consisted of National Colors and military weaponry. Symbolically, at the top was a rattlesnake with the inscription, "This We'll Defend." The possessive pronoun "This" is over a red Phrygian cap on a Liberty Pole.[17] This symbolically established liberty as the bedrock for the United States and religious freedom as its first right.

Religious freedom is the first freedom in the first amendment to the United States Constitution. The United States Army chaplaincy started one year before the signing of the Declaration of Independence in Philadelphia, Pennsylvania. In 1775, when Congress established the Army branches, the chaplaincy was the second branch, just after the Infantry. The Constitutional foundation of freedom and religious liberty stood in stark contrast with the totalitarian regimes of the Soviet Union, Nazi Germany, Fascist Italy, and Imperial Japan. Chaplain Arnold and the United States Army Chaplaincy underscored and emphasized religious liberty by supporting all faith groups. The Army chaplaincy reinforced Franklin Delano Roosevelt's January 6, 1941 State of the Union address, introducing the theme of the Four Freedoms that included the freedom of every person to worship God in his own way—everywhere in the world.[18]

A helpful tool to understand the context of 1937-1945 is the Operational Design Framework found in Joint Publication 5-0, *Joint Planning*.[19] Appendix 2 is an example of how OCCH could have used the Design Approach to define the problem and develop an operational approach to provide religious support during World War II. The framework will reveal how the mission of the chaplaincy has not changed. In addition, the example will show the function of Operational Design when applied with the current operational environment.

Chaplain (Major General) William Arnold

The man who led the Army Chaplaincy during World War II was Chaplain William Arnold, who was truly a man of firsts. He was the first Roman Catholic Priest to serve as Chief of Chaplains. He was the first chaplain to attain the rank of Brigadier General and subsequently Major General. He was the first chaplain to serve as a General Officer beyond the Chief of Chaplains position, as the Inspector General for Religious Affairs.[20] Chaplain Arnold was the only chaplain who ever recalled from retirement his predecessor, the former Chief of Chaplains Alva Brasted, to serve at the Fort Belvoir Chapel.

William Richard Arnold was born in Wooster, Ohio on June 10, 1881. His father, Augustine Adam Arnold, had Swiss heritage while his mother, Catherine Mary *nee Dalton,* was from Ireland. He went to the Saint Lawrence Roman Catholic School in Muncie, Indiana and then Saint Joseph's College, Rensselaer, Indiana. He attended Saint Bernard's Seminary in Rochester, New York, and ordained within the Rochester diocese, on June 13, 1908.[21] Chaplain Arnold was proud of his Swiss background, which went against the Irish Catholic dominance at the time in the United States Roman Catholic culture. He contended within his denomination over the heresy of Americanism—pronounced by Pope Leo XIII at the turn of the century toward American Roman Catholics who accepted the idea of separation of Church and State.[22] Chaplain Arnold remained faithful to his vocation within the Roman Catholic Church while also serving as an Army Chaplain. Chaplain Arnold's character and wisdom overcame prejudices within his denomination and by many Protestants.

Figure 2. Chaplain William Arnold. Source: U.S. Army Chaplain Museum.

Chaplain Arnold overcame Protestant bias and suspicion. Evidence of this suspicion ran deep, and this included Daniel A. Poling, a Protestant minister, Editor of the *Christian Herald*, and President of the World's Christian Endeavor Union. Daniel Poling was the father of Clark V. Poling, one of the four chaplains who died on the United States Army Transport ship the USAT Dorchester on February 3, 1943.

Some have argued that in 1960, Daniel Poling endorsed Richard M. Nixon for president, based on his distrust of John F. Kennedy's Catholic faith.[23] Poling was not entirely against Roman Catholics running for office, but struggled with their allegiance.[24] In Poling's own autobiography, he felt personally slighted because Congressman John F. Kennedy backed out of a formal banquet honoring the four chaplains at the Four Chaplains Chapel. Cardinal Dennis Dougherty, Archbishop of Philadelphia requested Kennedy not speak nor attend the function.[25] For Poling, this showed that unlike Arnold, Kennedy would place his allegiance of the Roman Catholic Church above the liberty of respecting the religion of other faith groups.

In Chaplain Arnold's Circular Letter 233, August 21, 1941, Arnold, unlike previous Protestant Chief of Chaplains, highlighted Daniel Poling's relationship with President Franklin Roosevelt by including a copy of a personal letter from the President to Poling. Throughout the war, Daniel Poling, as a correspondent, would carry messages from the President of the United States about the need to unite against totalitarianism. In addition, Chaplain Arnold would direct Daniel Poling, a Chaplain Major in the Army Reserves, to visit U.S. Army chaplains in Australia, China, North Africa, and Europe with messages from Chaplain Arnold, and report his observations to OCCH. This helped Chaplain Arnold gain awareness of religious support throughout the Army.[26] In February 1943, when word reached Chaplain Arnold that Poling's son was missing, he personally arranged for him to fly to England along the very route of the Dorchester with the hope of providing some solace to the grief stricken father.[27] Chaplain Arnold befriended Daniel Poling despite Poling's suspicions of the Roman Catholic Church, a tribute to Arnold's leadership skills and character.

Before his duties as Chief in December 1937, Chaplain Arnold's resume reflected many officers' experiences from his era. Arnold entered the Army in April 1913. Like President

Eisenhower, he did not see combat during World War I but served an assignment in the Philippines. After World War I, Arnold served numerous posts west of the Mississippi including Division Chaplain for the 1st Cavalry Division at Fort Bliss, Texas. In this assignment, he was the supervisory chaplain for the chaplains in the Civilian Conservation Corps (CCC) District that included Texas, Arizona, and New Mexico. Arnold was responsible for providing religious coverage to the 20 camps in that area. Due to the number of camps, Arnold would travel approximately 2,000 miles a month as part of his "circuit rides" and area coverage to provide religious support. Arnold requested additional support through the reserve component. Until then, he coordinated civilian clergy to care for the religious needs within the camps.[28] The struggles, experiences, and lessons learned while engaging with civilian clergy and governmental organizations providing religious support to personnel over a sizeable geographical area would help Arnold develop and implement his vision and plan during World War II. After his assignment with the 1st Cavalry Division, he then served six months as Director of the inactivated Army Chaplain School at Fort Leavenworth, Kansas. Chaplain Arnold persevered with the rest of the Army during the problematic interwar years.

An example of Chaplain Arnold's winsome ability and legacy occurred during 1936 in a train station in Texas. While assigned to Fort Bliss, Arnold was on one of his "circuit rides" visiting and providing Roman Catholic Mass and services for the many CCC camps in his area. At the train station, Chaplain Arnold noticed a fellow priest by the name of Terence Finnegan. Chaplain Arnold asked what he was doing in that part of Texas, and the man replied that he was in the area for his health. Chaplain Arnold then asked

Finnegan to join the reserves to help with serving the CCC camps. Finnegan replied he would have to get his bishop from his hometown of Hartford, Connecticut to release him. Arnold then spoke with his bishop who was a mutual friend and a decorated World War I chaplain, Chaplain Julius J. Babst. Babst persuaded Finnegan to join Chaplain Arnold at Fort Bliss.

When Arnold became Chief of Chaplains in 1937, he convinced Finnegan to pursue a Regular Army commission. However, Finnegan failed his physical for having only 19 out of 20 teeth. Arnold then approached General Marshall and the Surgeon General asking for an exception to policy. When they said it was up to Arnold, Arnold replied, "He's one of the best. I want him to serve the men as a chaplain, not to bite them." Finnegan received a Regular Army Commission and was assigned to Hawaii. Finnegan was with the 25th Infantry Division in Hawaii the day the Japanese attacked Pearl Harbor. Due to his performance in the Division, he became the Assistant Division Chaplain in 1942, and then the Division Chaplain in 1943. After the war, he transferred into the Air Force and attained the rank of major general and Chief of Chaplains of the Air Force in 1958.[29]

Chaplain Arnold's ability to influence extended to even the highest levels within his denomination. In August 1938, shortly after becoming Chief of Chaplains, a fellow priest and World War I veteran Chaplain Edmond J. Griffin went to visit the Pope in Rome. The purpose of this visit was to ask the Pope to approve a Vicar for the Military Ordinariate. Within the Roman Catholic denomination, the Military Ordinariate remains within the jurisdiction of the Roman Catholic Church and represents the Roman Catholic Church and their requirements and concerns to the government. Arnold wrote the following to Griffin, "Remember me to the Holy Father. Tell him the bit of purple he bestowed upon me has accomplished much for a good cause, but that I desire nothing more than his blessing on the task I am laboring at. Ask him to give us a Bishop with a manly character and a warm courageous heart, free from vanity, self-interest, and shyness."[30] Just over one year later, Bishop Francis Joseph Spellman became the Military Vicariate of the United States of America. In this capacity, Bishop Spellman supported Chaplain Arnold's vision of linking the Roman Catholic community with the urgent need to provide religious support to Roman Catholic parishioners serving in the Army.

Personally, Chaplain Arnold was a humble man who always remembered his modest beginning. Growing up he worked in his father's tobacco business as a tobacco stripper. A tedious job of removing the hard veins within tobacco leaves. Throughout his entire life, Arnold kept his union card from these years. Although a cigarette smoker, he hardly smoked in public. He enjoyed being around people and found ways to help them. The summer after his ordination, Arnold fulfilled a dream of his by becoming a clown for the Hagenbeck-Wallace circus during his summer vacation. As an Army chaplain, Arnold was more comfortable wearing his clerical outfit than his Army uniform. However, during his eight years as Chief of Chaplains, Chaplain Arnold would proudly wear his uniform along with other General Officers. On the War Department staff, Arnold was never shy about his role and responsibility. Arnold never abused his power, nor did anyone intimidate him from his mission to lead the Army Chaplaincy to its sacred calling of providing religious support to soldiers and their families.

Figure 3. General George C. Marshall awards the Distinguished Service Medal to Chaplain William Arnold. Source: U.S. Army Chaplain Museum. Edited to remove line in photo.

Chaplain Arnold remained faithful to his calling and faithful to the ideals of the Army Chaplaincy. He cooperated with those around him without compromising his integrity or his calling. Like his contemporaries Eisenhower and Marshall, Arnold demonstrated he could combine tact and diplomacy with leadership. Robert Gushwa's, *The Best of Times and the Worst of Times* contain examples of Arnold's leadership and appealing style:

> Bishop R. Bland Mitchell of the Protestant Episcopal Church in 1941 quoted someone as saying, 'If Chaplain Arnold had been in charge of things a few centuries ago there would have been no Reformation!' And the Southern Baptist magazine, said of him in November 1943, 'No man could be

fairer or more impartial in his dealing with men representing various religious bodies.'[31]

Donald F. Crosby's well researched work on Roman Catholic Chaplains, *Battlefield Chaplains: Catholic Priests in World War II*, interviewed many chaplains who served with Chaplain Arnold during World War II. Regarding chaplains helping one another, Chaplain Arnold once said, "that the battle within the Chaplain Corps between the different churches was almost as fierce as the one against the enemy." Crosby then writes, "His thoroughly challenging task was to keep the peace among the chaplains, an assignment that might well have broken a less resilient, resourceful, and patient man."[32] Although a victim of bias and prejudices, he did not turn to bitterness and revenge, but remained faithful to the ideals of his calling, seeking to lead and serve everyone under his care. His calling and chaplain identity resulted in the most significant transformation in Army Chaplain history. Chaplain Arnold utilized systems to transform the Chaplaincy in order to protect religious freedom and care for soldiers and their families.

Notes

[1] Elliott Vanveltner Converse, *Rearming for the Cold War, 1945-1960, History of Acquisition in the Department of Defense, v. 1,* (Washington, D.C. Historical Office, Office of the Secretary of Defense, 2012), 1.

[2] Arpi Dilanian and Matthew Howard, *Army Sustainment, The Number One Priority: An interview with Gen. Mark Milley*, (PB 700-19-02, Volume 51, Issue 2, April-June 2019), 10.

[3] https://www.rand.org/topics/military-transformation.html (accessed December 18, 2018).

[4] The author is thankful for Chaplain (Colonel) Retired, Duncan Baugh for his insight regarding the requirements for soldier readiness. The author is responsible for linking soldier readiness and requirements to DOTMLPF-P.

[5] https://8tharmy.korea.army.mil/site/assets/doc/mission/how-we-fight.pdf (accessed November 16, 2018).

[6] Captain James A. Moss, *Manual of Military Training*, (George Banta Publishing Company, Army and College Printers, Menasha Wisconsin, 1914). Colonel James A. Moss, Manual of Military Training (Second Revised Edition), (George Banta Publishing Company, Army and College Printers, Menasha Wisconsin, 1917), 18.

[7] Robert R. Venzke, *Confidence in Battle, Inspiration in Peace: The United States Army Chaplaincy 1945-1975 Volume V*, (Office of the Chief of Chaplains, Department of the Army, 1977), 41.

[8] https://readyandresilient.army.mil/ (accessed December 7, 2018)

[9] Defense Acquisition University, *Manual for the Operation of the Joint Capabilities Integration and*

Development System (JCIDS), February 12, 2015. https://dap.dau.mil/policy/Documents/2015/JCIDS_Manual_with_errata_through_20151218.pdf (accessed June 14, 2017).

[10] Larry A. Walker, ed., Department of the Army, *The Army Chaplaincy: Professional Bulletin Of The Unit Ministry Team*, PB 16-00-1, (Department of the Army, Summer-Fall 1999), 58-91.

[11] Department of the Army, *Transforming The US Army Chaplain Corps for the Modular Army*, TRADOC Pamphlet 525-1-05 Draft v.13, (Dept of the Army, TRADOC, Fort Monroe, VA, No Date, estimated date 2000), 1-37.

[12] Memorandum from OCCH to ASCC Chaplains, Subject: Establishment of Policy Concerning Structure and Functioning of Chaplaincy Participation in Process Integration Concept (PIC), (Department of the Army, OCCH, January 9, 2012).

[13] Pamela Wilson, ed., *The "Fifty-Six Mike"*, (IMCOM Religious Support Office, Volume 1, Issue 1, January 2014-Volume 10, October 2014).

[14] U.S. Department of the Army, Field Manual 6-0, Change 2, *Commanders and Staff Organization and Operations*, (Department of the Army, May 2014), para 16-43.

[15] Charleston, SC contained the largest Jewish population until the 1880 Eastern Europe Immigration to the United States.

[16] The National Security Act of 1947 changed the War Department Seal to the seal for the Department of the Army.

[17] The history of the Phrygian Cap and the Liberty Pole is symbolic of the freedom from slavery and tyranny.

[18] Michael Snape's recent and authoritative work *God and Uncle Sam: Religion and America's Armed Forces in*

World War II provides excellent insight into the social and cultural issues during 1937-1945.

[19] Chairman of the Joint Chiefs of Staff. *Joint Publication 5-0 Joint Planning*, (June 16, 2017), IV-7.

[20] Chaplain (BG) George Rixey served as Chaplain Arnold's deputy in OCCH. When Arnold served as an Assistant to the Inspector General, Chaplain Arnold took Rixey with him. When Arnold retired, Rixey retained his rank as BG and served as Arnold's replacement as Assistant to the Inspector General. Rixey's rank reverted back to Colonel when the position in the Inspector General Corps was abolished.

[21] William R. Arnold, "Message from our Chief of Chaplains," *The Army Chaplain: Official Publication of the Chaplains' Association of the Army of the United States, Vol VIII, No 3, (January 1938),* 82.

[22] Michael Snape,*God and Uncle Sam: Religion and America's Armed Forces during World War II, (*Boydell Press, 2015), 63. Conversation with Michael Snape, July 17, 2016.

[23] James Reichley *Religion in American Public Life,* (Washington, DC: Brookings Institution, 1985), 225.

[24] Daniel A. Poling, *Mine Eyes Have Seen: An Autobiography*, (McGraw-Hill, 1959), 185. In 1928, Poling praised Democratic Presidential Candidate Alfred E. Smith who was a Roman Catholic. He did this because Smith publicly pronounced his support of the separation of church and state and Poling "took him at his word"…that "if elected, he would be a President who happened to be a Catholic, and not a "Roman Catholic President.""

[25] Ibid., 256-261.

[26] Ibid., 206-211.

[27] Ibid., 202-203.

[28] Daniel B. Jorgensen, *The Service of Chaplains to Army Air Units, 1917-46*, (Wash, DC: Office, Chief of Air Force Chaplains, 1961), 52.

[29] Ibid., 53.

[30] Letter from Chaplain Arnold to Edmond J. Griffin, August 16, 1938, Quoted in Jorgensen, *The Service of Chaplains to Army Air Units, 1917-46*, 55.

[31] Robert L. Gushwa, *The Best And Worst of Times: The United States Army Chaplaincy 1920-1945*, (Volume IV, Office of the Chief of Chaplains, Department of the Army, 1977), 105.

[32] Donald F. Crosby, SJ, *Battlefield Chaplains: Catholic Priests in World War II*, (University Press of Kansas, 1994), xxii.

Chapter 2

Doctrine

The Army defines doctrine as "fundamental principles by which the military forces or elements thereof guide their actions in support of national objectives."[1] A suggested definition for religious support doctrine is, "Provide Soldiers and Families with, and advise leaders on, comprehensive religious support and care at every level of war in order to support national objectives and to uphold the free exercise of religion ensured by the First Amendment of the Constitution of the United States." There are three doctrine issues identified in this work—the free exercise of religion, chaplain professional duties, and graves registration.

Free Exercise

Starting in 1775, Army chaplains provided for the religious requirements for everyone within their organization. In the United States Code, Title 10, Armed Forces, "Each chaplain shall, when practicable, hold appropriate religious services at least once on each Sunday for the command to which he is assigned, and shall perform appropriate religious burial services for members of the Army who die while in that command."[2] A literal reading of this statute would imply that a chaplain only "works" one hour a week at a place of worship, and when tragedy strikes a unit. Initially and at first glance, what an Army chaplain does and how he performs his mission may appear to be insignificant and straightforward. However, when Title 10 begins as a starting point instead of an ending point, the role and functions of a chaplain change significantly and require a detailed explanation, so that chaplain activities nest within a commander's overall strategy and guidance.

Questions such as when is the best time to conduct a worship service, where should the worship service be located in order to maximize the greatest access for soldiers and their families, and what other religious faith group leaders

(RFGL) are required to meet the religious demographic profile of an organization. When tragedy strikes a unit, how does a chaplain provide support to the unit? When feelings and attitudes of revenge and depression surface after a tragic loss, chaplain activities to include care for the soldiers and advisement to the command is essential for the overall health and readiness of the command and its soldiers.[3] Additional questions that focus upon personal readiness such as financial stability, pregnancy considerations, substance abuse, and suicide awareness contribute to overall mission readiness. These questions are part of how a chaplain advises the command on morals and morale within a unit.

The National Defense Act of 1920 made this principle regulatory within the Army. In the first Army Regulation for Chaplains, Army Regulation 60-5, *Chaplains*, "Chaplains will serve as friends, counsellors, and guides, without discrimination, to all members of the command to which they are assigned, regardless of creed or sect, and will strive to promote morality, religion, and good order therein."[4] The doctrine of the chaplaincy providing for the free exercise of religion did not change during World War II. However, Chaplain Arnold and his staff in OCCH improved upon the requirement to care for all major faith groups. In addition, he corrected biases and prejudices by commanders and chaplains.

As in the First World War, OCCH reminded commanders and chaplains throughout World War II of the constitutional right to worship, or not worship, according to one's conscience. However, due to General John J. Pershing selecting the civilian Bishop Charles Brent to oversee religious support for the American Expeditionary Forces, there was no unified approach toward the free exercise of religion and correcting abuses. During Chaplain Arnold's tenure as Chief of Chaplains, some commanders and chaplains did not adhere to the free exercise of religion. Chaplain Arnold corrected this by exemplifying and publicly supporting all faith groups and correcting abuses.

On May 31, 1942, at Fort Belvoir, Virginia, soldiers received approximately 3,000 copies of Jewish, Catholic, and Protestant scriptures in "an impressive ceremony in the picturesque open air amphitheater on Sunday."[5] Inside the front cover of the Jewish, Catholic, and Protestant Bibles was a letter from President Franklin Roosevelt.

> To the Members of the Army:
> As Commander-in-Chief I take pleasure in commending the reading of the Bible to all who serve in the armed forces of the United States. Throughout the centuries men of many faiths and diverse origins have found in the Sacred Book words of wisdom, counsel and inspiration. It is a fountain of strength and now, as always, an aid to attaining the highest aspirations of the human soul.
>
> Very sincerely yours,
> Franklin D. Roosevelt.

Inside of the back cover of each respective Bible was a letter from Chaplain Arnold, asking if the soldier knew who their chaplain was.

Do You Know Your Chaplain?

> This copy of the scriptures should initiate and promote a warm friendship between you and your chaplain. He has studied the Word of God for years and uses it daily for his own strength and comfort, and for the instruction of others. His love for the Scriptures makes him your friend and guide. When he counsels you he speaks with the knowledge

and charity found in this little volume. A soldier who knows the Word of God and honestly tries to observe His laws is a man of power and influence among his fellows and exalts his military service to the high level of religious faith, courage and loyalty.

William R. Arnold
Chief of Chaplains[6]

Including his words with that of the President's sent a unified strategic message that Army Chaplains will not favor one faith group over another. Chaplains, regardless of their religious faith, should know the faith requirements of the soldiers within their organization. Anything counter to this does not support the President's vision of "many faiths and diverse origins…attaining the highest aspirations of the human soul."

Figure 4. Jewish, Roman Catholic, Protestant – Government Issued Scriptures
Source: The author's personal collection.

Shortly after the ceremony, an embarrassing situation arose when individuals complained that the Catholic New Testament Bible published by the Government Printing

Office contained anti-Semitic language. What made it personally humiliating for Chaplain Arnold was the wording in the book "Published under the direction of the Chief of Chaplains." The words in this reference went contrary to Arnold's personal belief. The anti-Semitic reference came from the editor, a Roman Catholic Priest, Father Joseph F. Stedman's note to the Apocalypse (Revelation 2:9): "The Jews are the synagogue of Satan. The True Synagogue is the Christian Church."[7] Arnold quickly stopped any further printing of this version. The Army then provided another version without the anti-Semitic reference, and a new Roman Catholic New Testament of the Gospels and the Apostles (Acts) that provided for the religious needs of the Catholic soldiers.

The incident within the military highlighted the bias and prejudices of Stedman. The anti-Semitic wording reflected a small and embarrassing sect within the Roman Catholic Church. The prompt action by Chaplain Arnold and the Army became a catalyst for reform for Stedman. This public rebuke changed Stedman and his writings for good. As early as August 1941, the Military Ordinariate received requests for a small missal but could not afford it at that time.[8] However, Stedman's work funded through the Confraternity of the Precious Blood of which he was the director remained the preferred aid to assist with the daily soldier's spiritual needs. The fact that his published missals, which came out after the anti-Semitic footnote, were free of any bias and prejudice and allowed by the military through to the Vietnam War, show that although Stedman was no longer alive, his writings continued to serve military personnel. The redemptive actions of the Army chaplaincy allowed Stedman to conform to the foundational principles of the free exercise of religion that prohibits the restriction of another religion through word or deed.

The European Theater of Operations After Action Report shows the immediate correction, the continued monitoring of religious materials, to include Stedman's Roman Catholic material, and the impact and purpose in providing religious support to soldiers.

> Two editions were offered for Roman Catholic use. In spite of the title on the cover, "New Testament, Roman Catholic Version," neither was a complete New Testament. The earlier edition was only a selection of daily readings based on a harmony of the four Gospels and the Epistles. The second edition includes only the four Gospels and the Acts of the Apostles. The consensus of Roman Catholic chaplains interviewed was that the Roman Catholic version should be a complete New Testament, with the same general physical characteristics as the Gideon edition of the Authorized Version New Testament. In greater demand among Roman Catholic troops than any edition of the Scriptures were denominationally issued prayer books and the Reverend Joseph P. Stedman's daily and Sunday missals.[9]

Chaplain Arnold and the government corrected the abuse and continued to monitor religious bias and prejudice in material reaching Army soldiers. It is important when contracting with religious organizations who assist with providing religious support, that the contractors fully understand the stipulation that when supporting one faith group, they cannot criticize or hinder another faith group. Chaplain Arnold's leadership reveals that he was quick to address and correct abuses and discrimination even within his own denomination.

There are, at times, zealous commanders or chaplains who feel it is their mission to force soldiers under their responsibility to attend or be exposed to religious services and events. Early in Chaplain Arnold's tenure, this abuse occurred within the Civilian Conservation Corps Camps and the Civilian Military Training Camps.

> Occasionally information comes to this office that in CCC camps and C.M.T. Camps men are compelled, against their wishes, by direct order or indirect pressure, to attend religious services under the guise of military or patriotic formations.
>
> Chaplains and other officers, no matter how praiseworthy their motives, must remember that in the Army and its various components there is no law, regulation, or custom which compels a man to attend a religious service contrary to his conviction or against his wishes.
>
> This respect for the individual's religious freedom also forbids the unethical use of any indirect or subtle pressure. The smaller the minority group the more need of care for its rights and privileges. Otherwise, religious fervor may be found in offense against real charity. The best religious technique is persuasion by example, not by coercion.[10]

The Civilian Conservation Corps Camps and the Civilian Military Training Camps were Depression-era programs that helped the military conduct mobilization and organizational exercises in spite of the economical constraints. The Army Chaplaincy utilized these unconventional organizations as a means to train and monitor for the free exercise of religion.

The respect for an individual's religious freedom also applies between chaplains. In Circular Letter number 8, the Military Ordinate for the Roman Catholic faith in the United States Army provided the same rationale for religious freedom among chaplains.

> "If a Protestant chaplain asks you to partake in a joint service (or a regimental commander orders you to do so) explain to him politely that this cannot be done. We have our pattern of religious worship, which is determined for us by the Sacred Congregation of Rites. Those who reject the authority of this Congregation (and of the other Sacred Congregations and Offices which rule the Catholic Church) cannot reasonably expect us to sanction their rejection. We recognize their right to worship God according to conscience, and we expect similar recognition – which is guaranteed by our Constitution."[11]

The leadership within the military and the agencies that send their chaplains into the military must continuously monitor and assess the mission to provide for the free exercise of religion. Moreover, leaders must immediately correct all abuses that hinder the free exercise of religion.

By October 1943, Chaplain Arnold received several complaints from various sources, Army and Civilian, about tracts and pamphlets in racks and on tables situated in chapel vestibules that attacked some religious groups, religious customs, or religious teachings. Chaplain Arnold replied,

> These "ANTI" writings, offensive to members of faiths toward which their sentences are directed,

are being placed with other literature acceptable to all,...It is desired that no literature be placed in any one of these racks which is offensive to any race or creed.[12]

Less than one year later, the Roman Catholic Military Ordinariate took issue with a prayer that promised that "whosoever reads it or wears it on him will never, burn, nor drown, will never be made a prisoner in war, nor ever be vanquished."[13] This example shows the importance of the denomination's responsibility to exhort their chaplains to monitor religious tracts and prayers that do not respect religious liberty and overall mental health of their soldiers.

During World War II, Chaplain Arnold and OCCH continued the proud tradition from 1775, supporting the doctrine of free exercise of religion and corrected abuses when they occurred. It is crucial for leaders, chaplains, and the denominations that send their chaplains into the Army to continually monitor and support the free exercise of religion and apply corrections when needed.

Extra Duties

In 1937, the Army had grown in personnel strength from 118,750 to 165,000 soldiers. However, the number of chaplains remained unchanged. Chaplain Alva Brasted acknowledged the issue of chaplains supervising recreational and athletic activities, and it is for this reason that Brasted stated a need for additional chaplains.[14] Many times chaplains spent more time with recreational activities than with religious duties. Brasted's approach to this problem was to increase the number of chaplains based upon a requirement that was not doctrinal nor supported the free exercise of religion. In contrast, Arnold used a doctrinal approach instead of a personnel approach to solve this problem and highlighted the priority of chaplains performing religious duties.

Before World War II, chaplains performed extra duties. Some of which, such as mess hall officers or trial counsel, conflicted with their professional duties. Soon after Chaplain Arnold's appointment as Chief of Chaplains, General Marshall asked if there was anything he could do to help chaplains. Chaplain Arnold replied, "protect them from secular duties." He went further and stated, "Doctors do not have to perform extraneous duties. This new Army will need religion." General Marshall issued an order limiting chaplains to their professional duties.[15]

Performing extra duties by chaplains was a matter of concern for Chaplain Arnold and many denominations. In the first eighteen months of Chaplain Arnold's career as a chaplain, he also served as a Post Exchange officer. During his duty, he once remarked to his Commander, "All I can do is see that everything gets into the till." The Commander replied, "That's what I want. The former officer got off with more than $2,000." Arnold was also a trial defense counsel for two years and during that time defended more than 80 men.[16] Arnold's experience influenced his views of chaplains performing duties that do not directly influence religious support.

During the interwar years, many denominations did not provide chaplains to the Army because of the extra duties. These extra duties included showing Hollywood movies on the Sabbath. This practice went against the moral beliefs of several fundamentalist chaplains. The 1926 and the 1937 edition of the Army chaplain manuals encouraged chaplains to show movies and to perform as recreation officers. When Chaplain Arnold became Chief of Chaplains and issued an updated version in 1941, chaplains no longer had to show movies on the Sabbath or perform as recreation officers. Arnold had these sections removed.[17]

Although General Marshall issued a General Order and a revised chaplain's manual removed recreational activities from the publication, Chaplain Arnold still had to intervene on occasion.[18] In Circular Letter 201, Feb 20, 1940, Chaplain Arnold personally was involved regarding a Commander who assigned his chaplain with extra duties as a recreation officer. Chaplain Arnold requested, not demanded, removing the chaplain from his extra duties:

> I recently had occasion to communicate with one of our chaplains concerning his assignment as Recreation Officer. The Commanding General saw the letter and made the following comment: "Since the receipt of your letter I have taken steps to conform with your request by relieving the Chaplain of these extra duties and detailing another member of my staff in his stead. I also wish to profit by this opportunity to express my appreciation of the fine quality of service which the Chaplain has rendered in this field as well as in that pertaining to his normal functions.[19]

Chaplain Arnold used diplomacy and tact with the chaplain's commanding officer. There may have been a reasonable explanation. The assignment as a recreation officer could have been only temporary until a new officer showed up at the unit. Regardless, Chaplain Arnold's interaction allowed for dialogue and furthered a solution to this problem.

Chaplains could support some recreational activities. The support included advisement to the Command and soldiers and the ministry of presence in these activities. On November

6, 1940, a directive from Chaplain Arnold to chaplains at reception and replacement centers provided wisdom and guidance regarding chaplain activities. At reception centers, selectees were there for a period of four days to four weeks. For many selectees, this was their first exposure to military life. The result was significant stress for the selectee. Chaplain Arnold provided advice and encouraged chaplains to pursue alternative recreational activities in four areas: religious, pastoral, cultural, and patriotic activities. Replacement center trainees could be there from four weeks to several months. Chaplains provided programs in greater detail in the same categories. For cultural activities, "the chaplain should not duplicate but cooperate with and supplement those of the Morale Division. He must be prepared to advise and counsel with the men as to the continuation of their reading and study, their hobbies and all legitimate activities not directly concerned with military training."[20] In the Technical Circular Number 4, March 1, 1942:

> Entertainments, sports, games, and the like are commonly a responsibility of other officers, but a wise chaplain will cooperate fully with all forms of wholesome recreation. He will seldom have time to organize and direct elaborate affairs of this type but may make suggestions which others will carry out successfully.[21]

Chaplains did not perform strictly secular duties that took them away from their professional duties. However, chaplain involvement in recreational activities provided moral encouragement to all soldiers who participated in them. The ministry of presence by chaplains in authorized functions is foundational for the Army Chaplaincy and essential for the morale and moral fabric on any organization.

Graves Registration

Nurturing the living, caring for the wounded, and honoring the fallen are the three competencies for the Army Chaplaincy. Honoring warriors who died in battle has a long tradition in military warfare. The impact Chaplains made in the development of honoring the fallen and incorporating it into their doctrine became an enduring legacy. Although an Army Chaplain started the process of honoring the fallen in the Spanish American War, it was World War II that presented the occasion to transition the program entirely to the Quartermaster Corps. Today, the Army Chaplaincy continues in the tradition of requiring every chaplain to provide religious rites and services throughout the entire process to honor the fallen.

Identification tags, commonly known among United States soldiers as dog tags, are reminders of the ultimate sacrifice soldiers make. Historically and in other countries, soldiers refer to dog tags as tombstones. In the United States military tradition, when a soldier dies, one dog tag remains with the personal belongings and another dog tag remains on the person of the deceased. Modern science, DNA, and technology may have made these identifications tags obsolete. However, their symbolism represents an important note in military history and specifically the chaplaincy. During the American Civil War, soldiers began to wear identification tags. Many historians agree that the American Civil War greatly influenced modern day warfare. Recently, historians began to focus upon how the United States during the Civil War honored their war dead.[22] During the war in the United States, death reached every community. Religion, and in particular Protestantism, at that time emphasized the rapture of the dead, which meant the physical raising of the body from the grave and the ascension to heaven.

Families on both sides of the war went to great lengths to protect and secure their fallen. After many battles, private embalming institutions embalmed select deceased soldiers. Primarily, soldiers or their extended family who could afford the service, made arrangements ahead of time in the event they became a casualty. The embalming institutions helped preserve the body, and indirectly, the memory of their fallen loved one. Chaplains supported this process by providing religious services in the field and at the place of internment.

Chaplains followed this tradition during the Spanish American War, and one such chaplain was Chaplain Charles C. Pierce, also known as the "Father of Mortuary Affairs." Pierce was born during the Civil War. After Pierce completed his civilian training, he served as an Army Chaplain and carried the importance of honoring the fallen to the Philippines, where he served in the Morgue and Identification Laboratory. During this assignment, Pierce focused upon preparing the bodies for return to the United States and created detailed record keeping of the process. At the start of World War I, Pierce came out of retirement, commissioned as a Major in the Quartermaster Corps, and directed by the War Department to establish Grave Registration as part of the Quartermaster Corps responsibility. Pierce's work influenced other nations in the establishment of their Graves Registration programs.[23]

The National Defense Act of 1920 and Army Regulations linked chaplains with Graves Registration. A survey of Army doctrinal publications and training from 1924 to 1948 will reveal a shift from the Chaplain Corps to the Quartermaster Corps for overall leadership and responsibility for graves registration. In 1948, this resulted in the Quartermaster Corps providing logistical and administrative support while the chaplains provide religious and ceremonial functions.

On February 15, 1924, the Army published the first edition of Army Regulation 60-5, *Chaplains.* Paragraph 13 contains the following for Graves Registration.

> Graves Registration.
> a. In time of war chaplains will often be charged with duties pertaining to graves registration personnel, particularly in the absence of such personnel with their respective commands. The functions of chaplains, when charged with burials, are so closely identified with those of the Graves Registration Service that they should familiarize themselves with such graves registration regulations as may be in force at the time in order to qualify themselves with these duties.
>
> b. Location of cemeteries.-Chaplains should familiarize themselves with the location of all authorized military cemeteries within the area of their respective commands, and when in charge of burial parties should, as far as possible, see that none but authorized cemeteries are used.
>
> c. Accuracy of records.-Chaplains in charge of burials will make careful comparison of their records with hospital, company, regimental, and divisional records whenever possible, in order that no discrepancies may occur as to the full name of the decedent, his organization, date of death and time and place of burial. Any error in an inscription or in the position of a grave marker which may come to the notice of chaplains at any time should be reported through channels to the commanding

> officer of the local grave registration unit or, in the absence of such commanding officer, to the nearest officer in charge of graves registration work. Chaplains will make no changes either in inscriptions or in the position of grave markers unless specifically ordered to do so.

The 1937 Army Regulation removed paragraphs b and c. The 1946 Army Regulation removed the Graves Registration section entirely, replacing it with the straightforward statement that "Chaplains are not available for detail as Army exchange, athletic, recreation, graves registration, welfare, morale, information-education, personal affairs, or special service officers."[24]

The incorporation of chaplain duties into the training of chaplains and chaplain training manuals also evolved during the interwar years.[25] During World War I, chaplains received a brief overview of graves registration at the chaplain school in France. In the morning, "several talks and instructions were given to them covering the identification of fallen soldiers, their burial, marking of the grave, position of graves, personal effects of the dead, notification of the Central Records Office and writing to the soldier's family."[26]

In January 1919, the Chaplain School in France closed. In 1928, the Chaplain School at Fort Leavenworth closed due to the lack of resources and enrollment. After the school closed, the Army Chaplaincy used correspondence courses to train new chaplains. However, due to the prioritization of course requirements and the lack of resources to oversee the correspondence courses, the curriculum between 1926 and 1937 did not include graves registration. In addition, the Director of the Chaplain School during this time also had the dual role of providing religious support to Fort Leavenworth.

As a result, OCCH oversaw the correspondence courses. When Chaplain Arnold became Director of the Chaplain School early in 1937, he influenced the Chief of Chaplains to include graves registration as part of the overall correspondence course for new chaplains. When the Chaplain School opened in 1942 at Fort Benjamin Harrison, an entire block of 15 hours out of the required 200 hours consisted of Graves Registration tasks and responsibilities.[27] Chaplain Arnold recognized a gap in training and made adjustments in order for chaplains to honor fallen American soldiers.

Similar to the establishment of a chaplain school in France during World War I, there was a Graves Registration School established in the United Kingdom during World War II. The European Theater of Operations After Action Report found that chaplains attendance at the Quartermaster School Graves Registration Course in the United Kingdom was "pointless."[28] The urgency of training basic tasks to get soldiers to the front superseded the quality of the training.

In 1926, a small chapter on the duties of chaplains for Graves Registration made it into the first official chaplain training manual.

> **Identification and burial of the dead-Graves registration**.-. *Existing orders*.-Since the chaplain is frequently charged with the burial of the dead it is his duty as early as possible after the outbreak of hostilities to familiarize himself thoroughly with all current orders and bulletins dealing with the burial of the dead, the disposition of the effects of the dead, and grave registration. The chaplain can also be helpful in this connection by frequently reminding both officers and men of the importance of always wearing the regulation means of identification.[29]

In the 1937 and the 1941 Technical Manual editions, the duties of chaplains for graves registration remained unchanged until the publication of the 1944 edition of the manual, which removed it almost entirely.

On September 23, 1941, the War Department published the first Graves Registration manual, Technical Manual 10-630, *Graves Registration.* In this manual, an entire chapter includes chaplain duties. The guidance echoed the chaplain's manual regarding overseeing the burial for the dead.

> In time of war chaplains will often be charged with duties pertaining to graves registration personnel, particularly in the absence of such personnel with their respective command…they [chaplains] should familiarize themselves with such graves registration regulations.[30]

The manual also directed chaplains to carry with them Graves Registration materials such as administrative documentation and materials to protect personal items of the deceased. On March 27, 1944, the War Department published Technical Bulletin 10-630-2, *Graves Registration.*[31] Technical Bulletins were a way leadership quickly provided information to soldiers in the field. This small 23-page document includes specific details about burials on the battlefield. Absent from this bulletin is any mention about chaplain duties. In January 1945, the War Department published Field Manual 10-63, *Graves Registration.*[32] This manual replaced the 1941 edition and copied almost verbatim the 1944 technical bulletin with several exceptions. Chaplains now had to annotate on the burial form the religious rites performed for the fallen soldier. Therefore, January 1945 is a significant date for graves registration and the Army chaplaincy. First, professional "career officers" from the Quartermaster Corps were now in the lead and responsible for the care of the remains of U.S. soldiers. Chaplains no longer led or assisted with the administrative portion of accounting for the remains. Second, the sacred function of administering the religious requirements for fallen soldiers and honoring the dead now became a core competency for the Army chaplaincy.

Chaplains were not Graves Registration officers. The Quartermaster Corps provided officers and soldiers for this function. However, doctrine within the manuals was not always clear. It appears the Quartermaster Corps was not able to fill the number of slots for Graves Registration Officers and relied upon many chaplains to fill these gaps.[33] One example shows how chaplains were trained and ready to provide support.

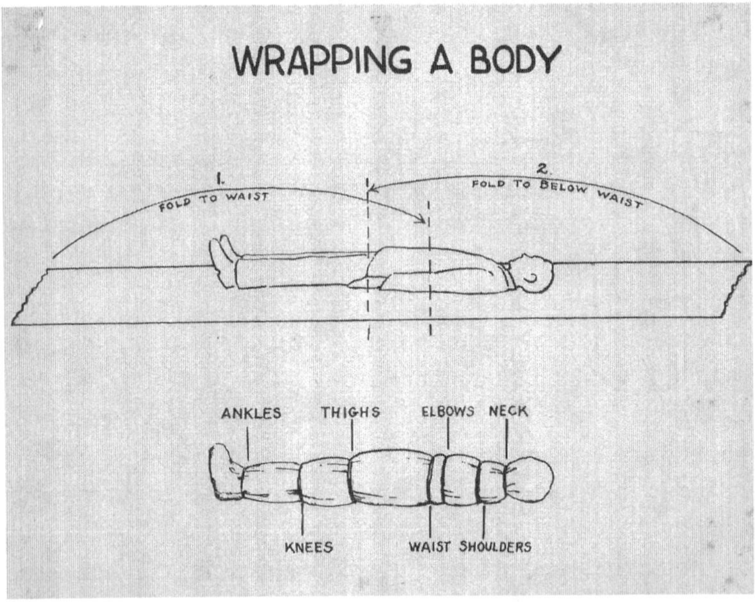

Figure 5. Chaplain School Graves Registration Instruction Handout On Wrapping The Body. Source: The author's personal collection.

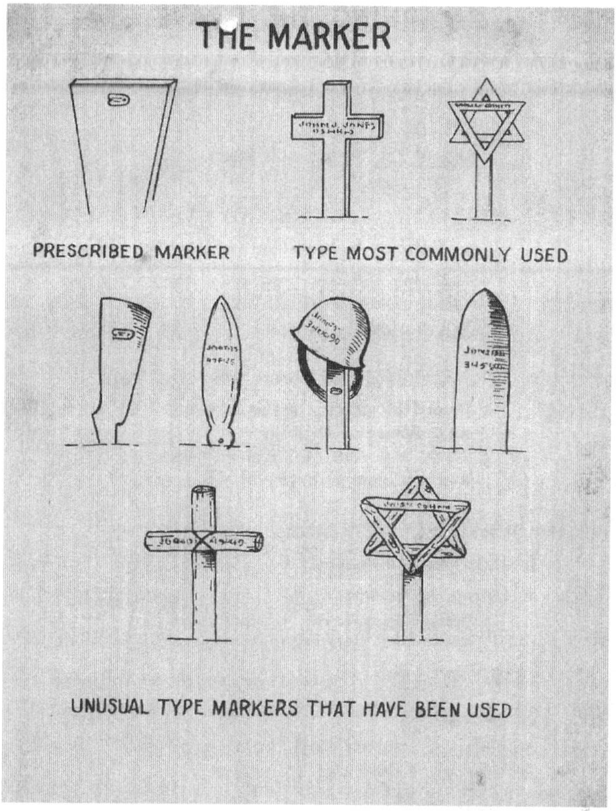

Figure 6. Chaplain School Graves Registration
Instruction Handout for Grave Markers
Source: The author's personal collection.

In September 1943, at the battle of Salerno, Chaplain Lawrence Hertzog was with his men of the 630th Anti-Aircraft Artillery (Automatic Weapons) Battalion. During the intense bombardment of the landing, Hertzog's boat reached shore, a distance away from the designated spot. Walking toward his unit, he noticed the graves registration team was struggling with the task of burying the large number of dead soldiers. The cause of the confusion was a lack of leadership. The appointed Graves Registration Officer could no longer function due to his wounds. Hertzog relied upon the guidance from the manuals and his training, and gathered the team and led them in their duty to properly account for the dead soldiers.[34]

In May 1946, Congress authorized the Secretary of War to return the remains of American men and women who died overseas between September 3, 1939 and July 1, 1946. The Quartermaster General established the American Graves Registration Service to conduct all administrative matters to the repatriation of remains.[35] With the professionalization of Graves Registration and the infusion of adequate personnel to perform the administrative duties, chaplains performed the sacred duties of honoring the fallen and caring for the next of kin. In 1946, the Office of the Quartermaster General for the War Department published a pamphlet *Disposition of World War II Armed Forces Dead* for those buried outside the United States. [36] This pamphlet provided information about how the next of kin could contact the War Department to inquire about the status or ministration of their loved ones. For Army personnel, they could go directly to the Chief of Chaplains, for the Navy, they would have to go through Navy Personnel for the religious ministrations performed for their loved ones. Chaplains provided religious services at national cemeteries when the family could not provide their own minister for the internment. Active duty chaplains would coordinate with National Guard or Reserve Chaplains, or with veterans' organizations to provide religious ministrations.[37] When the Quartermaster Corps completely took responsibility for Graves Registration and therefore the implied task of honoring the body, the Army Chaplain took this opportunity to expand upon a critical component of honoring the dead. The Army chaplaincy established direct and personal access that included care for the next of kin.

On July 1947, the War Department published Technical Manual 10-285, *Deceased Personnel*. This manual clarified the Quartermaster Corps role in record keeping and the disposition of the remains. Chaplains are "at the direction of the Installation Commander before shipment of the body" and "conducts funeral services when internment is in a post cemetery."[38] One month later, In August 1947, the Quartermaster Corps published Quartermaster Corps Manual 16-2, *Postwar Graves Registration Activities in an Overseas Theater* RESTRICTED. This document's only reference to a chaplain is on the last page, stating the Graves Registration mission is complete when the Chaplain conducts the graveside services.[39] The first official guidance for chaplains to perform the sacred duties of conducting a military funeral did not occur until War Department Pamphlet No 21-39, *Conduct of A Military Funeral*, September 1947.

During the entire war, chaplains would gather the dead, both friendly and enemy, place them on their jeeps and return them to the processing area. Caring for the dead was so important, that Chaplain Arnold noted in June 1945 in his public Circular Letter 297 "In the cases where American soldiers are buried in advance areas, these chaplains usually go out with the burial parties and conduct the services on the spot." There, chaplains would take turns performing "special duty" interning the remains and performing sacred services according to their faith. By June 1945, "the Army established forty cemeteries on the Continent, and each is visited at least once a day at regularly scheduled burial hours by Army chaplains of the Protestant, Catholic and Jewish faiths, expressly for the purpose of conducting burial services."[40]

At a cemetery, there would always be on rotational bases a Jewish, Roman Catholic, and Protestant Chaplain. These were extra duty assignments for chaplains. The European Theater After Action Report recommended one Roman Catholic and one Protestant Chaplain assigned to a quartermaster graves registration company. Although the number of Soldiers assigned to a graves registration company increased during World War II, the number never exceeded 200 soldiers.[41] Assignment to the grave registration company is far below the 1,200 prescribed number of soldiers for one chaplain. Also, the requirement of the mission drives the number of chaplains assigned to a unit.

An unfortunate incident occurred when a Jewish family received a letter from a Christian Chaplain stating that he performed proper services for Private Leonard Shapiro of the 80th Infantry Division, who died August 20, 1944 in France. The family was horrified, and on December 5, 1944 notified OCCH and in particular Chaplain Aryeh Lev, the Jewish Chaplain serving in OCCH.[42] Chaplain Lev gave the information to Chaplain Arnold and quickly looked into the matter, discovering the soldier's record did not indicate him as Jewish.[43] Chaplain Arnold apologized to the family, and in June 1945 wrote the following in his Circular Letter for the public:

> Every effort is being made by the Army to transfer bodies from these isolated graves to the established cemeteries as rapidly as possible, and if they have not received the burial service of their own faith the first time, they are certain to when they are reinterred.[44]

Another situation similar to this was the death of the Third Armored Division Commander, Major General Maurice Rose. Rose came from a devout Jewish family. He was a combat veteran of World War I and served during the interwar years. Rose married Virginia Barringer, the daughter of Major James Barringer, a Christian family.[45] During World War II, Rose developed a close relationship with his division chaplain, Chaplain (Lieutenant Colonel) Paul Maurer, a Christian Congregationalist. On November

12, 1944, Maurer preached a sermon[46] that Rose liked so much; he had it lithographed and sent to all the soldiers throughout the division.[47] On March 30, 1945, General Rose died by enemy fire. His Division Chaplain provided the burial service and placed a Christian cross upon his grave. His Jewish family protested only to find out from a Jewish army chaplain, Chaplain David Lefkowitz from VII Corps, who provided Jewish services for 3rd Armored Division, that he knew General Rose and that Rose converted to Christianity and therefore received the proper burial according to his Christian faith.[48] Despite some difficulties and sensitive issues, this is an excellent example of chaplains from different faith backgrounds working together to accommodate and honor the faith of all fallen soldiers.

There is one account that best represents how chaplains respected the faith of the fallen soldiers and highlights the logistical effort required to honor their belief. On May 13, 1945, a plane crashed with 21 people on board in the remote mountains of New Guinea. Chaplains were unable to reach the site. However, Chaplain Augustus Gearhard, a Roman Catholic Priest, Protestant Chaplain Carl Mellberg, and an Army Air Force crew flew to the site on May 29, 1945, and dropped crosses and Star of David grave markers to Filipino paratroopers who made it to the crash site. Then while flying overhead, each chaplain recited the funeral rituals for each of the victims. Because no Jewish Chaplain was available within a thousand miles, Chaplain Mellberg recited the Mourner's Kaddish for the Jewish woman of the Women's Army Corps who died in the crash.[49]

Chaplains who protect the free exercise of religion and stay focused on their sacred duties sustain their chaplain identity. Supporting and protecting soldiers' freedom to worship is an essential task and duty for every chaplain. Engaging and contributing to recreational activities reveals the incarnate ministry of chaplains that transcends the earthly boundaries of race, creed, or gender. Chaplains who protect the free exercise of religion, to include the sacred duties of honoring their death and providing religious and pastoral care to the next of kin, reflect their calling and identity as religious and spiritual leaders.

Notes

[1] U.S. Department of the Army, Army Regulation 5-22, *The Army Force Modernization Proponent System*, (Washington, DC: U.S. Department of the Army, October 28, 2015), 13.

[2] US Code Title 10, subsection 3547, page 1580 https://www.govinfo.gov/content/pkg/USCODE-2006-title10/pdf/USCODE-2006-title10.pdf (accessed February 14, 2018).

[3] Army pilot Warrant Officer Hugh Thompson Jr. who first reported the My Lai massacre was first cared for and advised by an Army Chaplain Carl E. Creswell to report the incident to the Division.

[4] U.S. War Department, Army Regulation 60-5, *Chaplains*, (U.S. Government Printing Office, February 15, 1924), 3.

[5] Headquarters Army Service Forces, Circular Letter 253, Office of the Chief of Chaplains, June 15, 1942.

[6] *Readings From The Holy Scriptures Prepared For Use Of Jewish Personnel Of The Army Of The United States*, (Washington, DC: Government Printing Office, 1942). *My Daily Reading from the Four Gospels and the New Testament: Gospels Unified, Epistles Unified Prepared For Use Of Catholic Personnel Of The Army Of The United States* (Washington, DC: Government Printing Office, 1942). *The New Testament Of Our Lord And Saviour [sic] Jesus Christ Prepared For Use Of Protestant Personnel Of The Army Of The United States*, (Washington, DC: Government Printing Office, 1942).

[7] *My Daily Reading from the Four Gospels and the New Testament: Gospels Unified, Epistles Unified* (Washington, DC: Government Printing Office, 1942), 559.

See Kevin L. Walters *Beyond The Battle: Religion and American Troops in World War II* (University Of Kentucky, 2013), 86-87.

[8] Military Ordinariate, Circular Letter 13, (462 Madison Ave NY, NY, August 25, 1941), 1.

[9] *Report On The Army Chaplain In The European Theater* (The General Board, United States Forces, European Theater, Study Number 68), 107.

[10] Headquarters Army Service Forces, Circular Letter 190, Office of the Chief of Chaplains, August 11, 1938.

[11] Military Ordinariate, Circular Letter 8, (May 26, 1941), 1.

[12] Circular Letter 277, Office of the Chief of Chaplains, October 1, 1943.

[13] Military Ordinariate, Circular Letter 37, (June 1, 1944), 2.

[14] Alva J. Brasted, "Chief of Chaplains Report," The Army Chaplain: Official Publication of the Chaplains' Association of the Army of the United States, Vol VIII, No 3 (January 1938), 99.

[15] Daniel B. Jorgensen, *The Service of Chaplains to Army Air Units, 1917-46*, (Wash, DC: Office, Chief of Air Force Chaplains, 1961), 87.

[16] Daniel B. Jorgensen, *The Service of Chaplains to Army Air Units, 1917-46*, 9.

[17] Robert Nay, "The Operational, Social, and Religious Influences Upon the Army Chaplain Field Manual, 1926-1952" (Master of Military Arts and Science thesis, Command and General Staff College, 2008).

[18] Ibid.

[19] Circular Letter 201, Office of the Chief of Chaplains, Feb 20, 1940.

[20] Headquarters Army Service Forces, *Directive for Chaplain Activities: Reception and Replacement Centers*, (Office of the Chief of Chaplains, November 6, 1940).

[21] Headquarters Army Service Forces, *The Regimental or Unit Chaplain*, Technical Circular Number 4. (Office of the Chief of Chaplains, March 1, 1942), 7.

[22] Drew Gilpin Faust, *This Republic of Suffering: Death and the American Civil War*, (Knopf, January 8, 2008).

[23] Dr. Leo P. Hirrel, Army Sustainment, *The Beginnings of the Quartermaster Graves Registration Service*, (PB 700-14-04, Volume 46, Issue 4, July-August 2014), 64-67.

[24] U.S. War Department, Chaplains, Army Regulation 60-5, *Chaplains*, (War Department, Washington D.C. December 12, 1946), paragraph 6g.

[25] The 1926 and 1937 manual titles were *Training Manual 2270-5*. The 1941, 1944 and 1947 manual titles were *Technical Manual 16-205*. The 1952 Manual title was *Field Manual 16-5*. All served the same purpose.

[26] Headquarters Army Service Forces, *History of USACHS: France 1918-Havard University* (Unpublished document prior to the establishment of USACHCS, original paper copy with Chaplain (Colonel) Robert Nay).

[27] Ibid.

[28] U.S. Forces, European Theater, *Report on the Army Chaplain in the European Theater,* Chapter 2.

[29] U.S. War Department, Technical Manual 2270-5, *The Chaplain: His place and duties*, (Government Printing Office, 1926), Para. 102.

[30] U.S. War Department, Technical Manual 10-630, *Graves Registration*, (War Department, September 23, 1941), 15.

[31] U.S. War Department, Technical Bulletin 10-630-2: *Graves Registration*, (War Department, March 27, 1944).

[32] U.S. War Department, Field Manual 10-63: *Graves Registration*, (War Department, January 1945).

[33] Jorgensen, *The Service of Chaplains to Army Air Units, 1917-46*, 191.

[34] *American Chaplains' of the Fifth Army*. (Pizzi and Pizzio, Milan, Italy, 1945), 15.

[35] Headquarters, Sixth Army, *Repatriation of World War II Dead*, (UTAFP-243, nd), 1.

[36] U.S. War Department: Offi ce of the Quartermaster General, *Disposition of World War II Armed Forces Dead*, (Government Printing Office, 1946).

[37] *Repatriation of World War II Dead*, 2.

[38] U.S. War Department, Technical Manual 10-285 *Deceased Personnel*, (War Department, July, 1947), 6.

[39] U.S. War Department, Quartermaster Corps Manual 16-2, *Postwar Graves Registration Activities in an Oversea Theater* RESTRICTED, (War Department, August, 1947), 103.

[40] Circular Letter 297, Office of the Chief of Chaplains, June 1, 1945.

[41] T/O 10-297, dated 1 January 1940; and T/O 10-297, dated 1 July 1943.

[42] Ronit Y. Stahl, *Enlisting Faith: How the Military Chaplaincy Shaped Religion and State in Modern America.* (Harvard University Press, 2017), 1-2.

[43] Many Jewish soldiers did not self-identify as Jewish in the European Theater.

[44] Circular Letter 297, Office of the Chief of Chaplains, June 1, 1945.

[45] Steven L. Ossad and Don R. Marsh, *Major General Maurice Rose: World War II's Greatest Forgotten Commander*, (Taylor Trade Publishing, 2003), 100.

[46] The sermon is not sectarian and includes references to crosses and stars of David over the graves at Normandy. Maurer also spoke prayers in English, Jewish and Latin "I Am The Resurrection And The Life…Sh' Ma Yisreayl… Requiescant in Pace."

[47] Headquarters, Third Armored Division, Letter from Major General Maurice Rose with sermon by Chaplain (LTC) Paul H. Maurer, November 13, 1944. Original lithographed paper with Chaplain (Colonel) Robert Nay.

[48] Steven L. Ossad and Don R. Marsh, 356-357.

[49] Army Forces Pacific, *History of Chaplains' Activities in the Pacific*, (Chaplain Section, Ground Headquarters, Army Forces Pacific, 1946), 283-284.

Chapter 3

Organization

Chaplain Arnold made three significant contributions to the organization of the chaplaincy. The first and most important contribution was adapting the size and function of the Office of the Chief of Chaplains (OCCH). The other success was establishing the foundation and organization for the future U.S. Air Force Chaplaincy. Third, an organizational structure that provided chaplains a ministry of presence to soldiers regardless of their proximity to the front line of battle. The Army defines an organization as:

> A unit or element with varied functions enabled by a structure through which individuals cooperate systematically to accomplish a common mission and directly provide or support warfighting capabilities. Subordinate units/elements coordinate with other units/elements and, as a whole, enable the higher-level unit/element to accomplish its mission. This includes the manpower (military, civilian, and contractor support) required to operate, sustain, and reconstitute warfighting capabilities.[1]

A suggested religious support definition for an organization is:

> Chaplain Sections and Unit Ministry Teams with capabilities and competencies integrated at every echelon, executing religious support functions, providing Unit, Area, and Denominational coverage, accountable to their assigned chain of commands and the chaplain technical staff channels to the Army Chief of Chaplains.

Chaplain Arnold's organizational changes strategically grew the capability of the chaplaincy and affected every level of war.

Office of the Chief of Chaplains

In order to integrate chaplain activities into the entire Army, OCCH reorganized the personnel and functions of its office. Four areas of emphasis contributed to the success of this restructure. First, Chaplain Arnold surrounded himself with competent staff officers. Second, Chaplain Arnold reduced the number of sections within OCCH. Third, Chaplain Arnold stressed the importance of the monthly reports that provided him an awareness of religious support at the tactical level and how supervisory chaplains were engaging their chaplains. Finally, he emphasized the importance of highlighting the ministry from the monthly reports through publicity and strategic communication.

Figure 7. Meeting of the Office of Chief of Chaplains, 1940 with a women's organization. LTC Goodyear (sitting second from left), Chaplain Honeywell (standing left), Chaplain Arnold (standing second from right), Chaplain Rixey (standing right). Source: Army Chaplain Museum.

A key individual who Chaplain Arnold utilized and was instrumental in synchronizing religious support within the Army during World War II was not a chaplain, but the Executive Assistant, Lieutenant Colonel Augustus S. Goodyear. Goodyear's training background was that of a lawyer; however, his branch in the Army was the Adjutant General Corps. Since the beginning of OCCH in 1920, Goodyear's tasks consisted of simple office administration.

In 1920, Chaplain Julian E. Yates performed as the Executive and did not utilize Goodyear's talents. Yates was unique as a chaplain for his strategic thinking and his organizational abilities. Yates oversaw the first chaplain's field manual published in 1926 and assisted with the first official Army and Navy hymnal.[2] After Yates, the position was underutilized, and Goodyear's abilities remained unappreciated. When Chaplain Arnold became Chief in 1937, Arnold gave Goodyear the freedom to manage the Corps.

Goodyear immediately went to work. In a speech to a chaplain's conference published in *The Army Chaplain*, Goodyear recommended:

> I have been connected with chaplains and their activities since the office was established in 1920, and often I have to refer to the Bible for that corps. I would suggest that you gentlemen make a very intimate acquaintance with TM 2270-5, entitled "The Chaplain." It has been recently been revised and I am sure we all would benefit by its perusal.[3]

Chaplain Arnold trusted Goodyear to manage and integrate the Army chaplaincy into the Army. He was able to do this because Goodyear did things the Army way. An example of this occurred the first year of Chaplain Arnold's tenure as Chief of Chaplains. Once a year, OCCH had to pay $5.00 for a bond for the Chief of Chaplains to validate the timecards for everyone in the office. In 1938, the time came to renew the bond. Goodyear suggested Chaplain George Rixey, Arnold's deputy, validate the timecards along with Arnold. No other chaplains in OCCH concurred because of the desire to save $5.00. When the routing slip reached Arnold's desk, Goodyear was the only one who supported having Rixey bonded. Arnold exercised tact, diplomacy, and supported Goodyear's request. Arnold mentioned how it would be nice to save $5.00 but then asked the staff how they could get paid if he were not in the office to validate the timecards. Goodyear's leadership, actions, and character influenced OCCH, the Pentagon, and the Army. OCCH functioned so well that General Marshall said that OCCH was the best run office in the entire War Department.[4]

LTC Goodyear's knowledge and ability facilitated Chaplain Arnold to have direct access to the General Marshall despite the levels of organization and bureaucracy. OCCH worked under the Commander for Army Service Forces Director of Personnel, who in turn operated under the Commanding General Brehon B. Somervell, Army Service Forces. This chain of command meant the chaplaincy was one level lower than the other Service Chiefs, and like every other General Staff and Staff function down to the brigade level, the Chaplain had no representation at the command and staff level. Despite this challenge and handicap, Army chaplains continued to work together within the military system and function with service members at every echelon of war, from remote arsenals and camps in the United States to the air and distant battlefields throughout World War II. See Appendix 3 "How The Chaplaincy Fits Into The Army Structure."

The contribution of having Goodyear, a skilled staff officer outside of the chaplain corps, to assist with managing and integrating religious support into the military contributed to the success of the chaplaincy during World War II. His action and reputation went beyond the Army. The Air Force Chaplaincy noticed the importance of having a skilled staff officer and selected a non-chaplain, Major Edward F. Donnelly as the Chief Executive Office for the newly established OCCH for the Air Force in 1948.[5] In recognition of his service, Goodyear became an honorary member of the Chaplain's Association. Even in death, Goodyear's grave is just a few feet from Chaplain's Hill in Arlington

Cemetery. Goodyear was the essential link that provided the professionalism and technical expertise that harnessed Chaplain Arnold's strategic leadership and vision to provide religious support.

Another key individual who enhanced the effectiveness within OCCH was Chaplain (Major) Roy J. Honeywell. Honeywell served in OCCH, Plans and Training Division starting on December 15, 1941.[6] Honeywell was a United Methodist Church minister from the Troy Conference (Albany, New York) of the United Methodist Church. Honeywell was an Army Reserve Chaplain who joined the Chaplaincy at the close of World War I. He was a full-time professor at Boston University's history department between 1920 and 1933. In 1941, OCCH mobilized Honeywell from the Reserves and appointed him to OCCH.

As an educator and a reservist, Honeywell understood the importance and the challenge of transforming civilian clergy into military chaplains. Honeywell's method included providing written information on the best ways to provide religious support. Many of the early publications from OCCH came from Honeywell. Honeywell also wrote the 1944 *Technical Manual 16-205, The Chaplain*. This manual was the first rewritten manual since 1926. The 1944 chaplain's manual is an excellent example of a wartime manual. Although not perfect, the author took into account civilian clergy who would be reading the manual and the questions they would ask regarding how to provide religious support in a military context. The manual is precise yet practical, realistic yet encouraging, earthly, yet a spiritual reminder of the higher calling chaplains have in the military. After he wrote the 1944 manual, he spent a short period of time at the chaplain school at Fort Devens, Massachusetts, and then returned to OCCH until his retirement from the Chaplaincy in 1948.[7]

In 1958, Honeywell, in his retired status, wrote the first official "blue book" history from OCCH, *Chaplains of the United States Army*.[8] It was in Honeywell's history of the chaplaincy where he wrote what could be an autobiographical account of the impact OCCH had during the war.

> During the gloomy days of 1942, morning devotions were instituted in the office of the Chief of Chaplains. Those members of the staff who chose to do so joined in 5 minutes of reverent supplication for the strengthening of their own spirits and for the welfare of all chaplains. Men on the distant battle lines knew the hour at which they were being especially remembered with confidence and affection among the immediate associates of their chief. This practice won the warm approval of the Secretary of War. The time was changed to noon and extended to 15 minutes, and workers in adjacent offices were invited to join. A considerable number regularly used a part of their noon hour in this way, and many hundreds participated year by year in special services for such occasions as Christmas and Lent.[9]

This account highlights Honeywell's character and example that refutes the false dichotomy that good staff officers cannot provide religious and spiritual support. He never forgot his calling to pastor souls. Even in the administrative role at OCCH, he had the unique ability to connect the disciplined requirements of military life with the duties of chaplains that often transcends human limitations and boundaries.

Another way Chaplain Arnold contributed to the organization and transformation of the Army chaplaincy during World War II, was by significantly changing the number of sections within OCCH between World War I and World War II. This reorganization resulted in efficient and monitored religious support throughout World War II.

From 1920 to 1939, OCCH operated with three chaplains, and three soldiers from the Adjutant General's office.[10] During World War II, the staffing grew to 26 chaplains and officers, to include two Warrant Officers from the Adjutant General Corps plus 125 Civilian clerks.[11] Before 1937, OCCH had ten sections, whose function was not well coordinated. With no Deputy Chief of Chaplains or functioning Executive Officer, the Chaplaincy had only minimal ability to fulfill its role as a proponent. When Chaplain Arnold became Chief of Chaplains, Chaplain Arnold and Lieutenant Colonel Goodyear reorganized the office and reflected how General George Marshall reorganized his staff. Namely by minimizing functions into a single department director. The result was a decreased number of individuals who competed for General Marshall's and Chaplain Arnold's time and resources. Chaplain Arnold and Lieutenant Colonel Goodyear combined the staff into the following four sections: Chief of Chaplains, Executive Division (Executive Assistant Lieutenant Colonel Goodyear and the Deputy Chief of Chaplains), Personnel Division, Planning, and Training Division, Statistics and Publicity Division. By March 1941, OCCH adjusted again to reflect the needs of the Army due to the number of maneuver exercises conducted in the United States and the newfound importance of the Army Air Corps. The new organization was as follows: Personnel Division, Ground Liaison Division, Air Liaison Division, Technical and Information Division, Administrative Division, Plans and Training Division, and Miscellaneous Division. Now, many of the other smaller sections were under one department head. This increased efficiency by condensing the functions within the office, allowed Chaplain Arnold to monitor key strategies and initiatives with his Division Chiefs.

One way that Chaplain Arnold and his team at OCCH monitored religious support within the organization of the chaplaincy was through the required monthly report. Situational awareness through reporting and communication is essential within any large organization. These reports allowed Chaplain Arnold to assess two key aspects of religious support: the provision of the free exercise of religion, and the character of the chaplain and the supervisory chaplain. An entire section of OCCH was responsible for gathering information from the monthly reports and compiling them into analytical data. Many chaplain files at the National Archives contain corrected reports from supervisory chaplains in the field. These reports contain corrections and added annotations for a specific ministry event. Supervisory chaplains followed the guidance from OCCH to monitor subordinate chaplains, and the reports helped to ensure effective religious support. This information also gave Chaplain Arnold situational awareness and accurate information of the status and wellbeing the chaplains within the Army.

One way to observe the impact of the monthly report is to use a modified DOTMLPF-P "Gap Analysis" to assess transformation. "Gap Analysis" compares actions before the change and actions after the change. In the World War I monthly report the chaplain was required to account for the number of births, baptisms, marriages, and deaths during the month. Another section included remarks on the moral condition and general condition of the post or regiment.[12] The questions within this report did not provide clarity and fidelity to the Chief of Chaplains regarding the activities by chaplains to ensure the free exercise of religion for all faith groups. A complete revision of the monthly report provided the required information for Chaplain Arnold.

There were several editions of the World War II monthly report. OCCH removed C.M.T.C. [Citizens' Military Training Camp, ed.] and R.O.T.C. [Reserve Officer Training Corps, ed.] organization. The following contains the information required in the monthly report.

- Record of births, baptisms, marriages, and deaths during the month
- List the number of services performed on a Sunday and during the weekdays
- What type of service they led
- The Chaplain must justify if he did not perform any services on Sunday
- How many services the chaplain participated in with other chaplains
- Number of services the chaplains attended
- Number of pastoral, educational, and recreational activities
- List all their activities in the civilian community
- Number of services conducted by visiting clergymen
- List the religious preference profile, broken down into Protestant, Catholic, Jewish, and Non-Denominational
- List any secular duties the chaplain was assigned
- Facilities provided for ministry and the Assistant who served with the chaplain.
- At the bottom of the report, the chaplain was "privileged to write directly to the Chief of Chaplains for information, guidance, or professional instruction, under paragraph 2a(4), A.R. 60-5" (ed, *The Chaplain* regulation).[13]

These reports allowed Chaplain Arnold to assess two key aspects of religious support: the provision of the free exercise of religion, and the character of the chaplain and the supervisory chaplain. In Circular Letter 248, April 1, 1942, the Chief of Chaplains captured the words of the V Corps Chaplain for himself and said the report was a "Character Reference" for the chaplain submitting the report. In the Circular, he writes:

> V Army Corps [Chaplain], states "The Chaplain who is turning in a careless and imperfect monthly report is turning in a careless and imperfect character reference for himself. Every time he does this he touches deceit, an unseen thing rising like a subtle essence to poison his future."

Interestingly, this comment indicated that the Chief of Chaplains looked at the report as a reflection of the character of the chaplain. Chaplains professionally performing their duties in every detail were very important to Chaplain Arnold. Second, Chaplain Arnold exalts the V Corps Chaplain as an example for all supervisory chaplains. As a supervisory Chaplain, he enforced the standards.

Chaplain Arnold believed that a chaplain's character must also apply to how chaplains interact to serve one another. In the 1943 publication, *The Chaplain Serves*, "The Supervisory chaplain does much to assist his colleagues by advising or encouraging when necessary, and by working out the highest degree of cooperation among them."[14] As the moral and spiritual leader in the unit, chaplains would sometimes get discouraged. The challenges of wartime ministry and the discouragement that comes with being the spiritual leader, it became imperative that chaplains support and encourage other chaplains. The character of the chaplain is essential as the spiritual and moral leader of the unit. In addition to providing for the free exercise of religion, the purpose was the moral and spiritual readiness of the unit. It begins with the chaplain and ends with the soldier. Arnold identified this concept during World War II. After the War, the Chaplaincy expanded upon this concept in several published documents.[15]

In December 1948, the Chaplain School published *Religious and Spiritual Morale Factors*. The Chaplain School republished it again in August 1950, and incorporated into the curriculum as *The Chaplain and Military Morale, ST 16-155*.[16] This in-depth document outlines measurable physical, mental, and moral factors of soldiers that contribute to morale and readiness. Much of this same instruction became part of *Chaplain's Hour*, monthly publications that transformed into the Character Guidance Lectures and Program. This instruction material is an excellent example of how supervisory chaplains correcting, clarifying, leading, and providing resources for the chaplain in the field.

In addition to this new curriculum, the Chaplain School also created *Chaplain Supervision: General Considerations ST 16-153*. This document shows that effective supervision learned from World War II included "spiritual, moral, and administrative leadership, combined with careful and coordinated planning, complete dissemination of information, the encouraging and helpful attitude of a good counsel, and a knowledge of the total situation based on a continuous program of fact-finding."[17] The messages significantly addressed the directive from General Marshall to the Chief of Chaplains to develop and care for the spiritual and moral readiness of soldiers. For in the mind of the leadership, the two were inseparable. When the people of the nation send their loved ones into the military, it became imperative to ensure the military would do everything in its power to protect them from spiritual and moral decay.

A way to demonstrate to the nation and the people how the military and in particular, the Army chaplaincy took care of their loved ones was through publicity. The monthly reports provided critical information for Chaplain Arnold to show in his circular letters to the nation the details how the Army was taking care of their sons and daughters in the military. The military had a reputation of attracting individuals with questionable character traits who would corrupt naïve individuals who joined the service. The chaplains published a monthly circular letter for the public to read that contained statistics and stories about ministry opportunities and testimonies from commanders, soldiers, and chaplains.

Publicity to the public assured that men and women in the Army received the spiritual care they needed for their welfare. The Chief of Chaplains welcomed "snapshots of chaplains' work, activities of religious organizations functioning in the Army or CCC" in order to assist with the publicity of chaplain ministry.[18] Public displays at conferences, state fairs, and department store windows supported the war bond drive, but also assured the public the provision of religious freedom was essential to the leadership of the Army.[19] Another purpose for the publicity was the "hope each chaplain will be challenged by this task" to assist with the recruitment of chaplains into the Army. Publicity and chaplain involvement for January 1943 contributed a net gain of 70 chaplains per week.[20]

Publicizing chaplain ministries greatly affected the success of the chaplaincy and recruiting efforts during World War II. Denominations within The General Commission on Chaplains had numerous publications highlighting chaplain ministry. Radio also helped to tell the chaplain story. Chaplain Arnold had the *Soldiers of God*, the Official Chaplain March song broadcasted on the radio during the Easter season. Jewish Chaplains had their program called "Message of Israel."[21] In 1942, the Army chaplaincy hosted the daily radio program "Chaplain Jim – U.S.A." Chaplains throughout the Army contributed material for this program.[22] The Chief of Chaplains personally sent certificates of appreciation to synagogues and churches who supported the Army Chaplaincy.[23] Department store windows or town fairs contained traveling displays of chaplain equipment.[24]In Circular Number 268, February 15, 1943, Chaplain Arnold wrote: "This favorable publicity now being given chaplains' work make this time propitious for active procurement and the Chief of Service hopes each chaplain will be challenged by this task."[25]

The history of the film, *The Army Chaplain*, shows the importance of having a professional publicity campaign. In 1942, Pathe News filmed the chaplain school under the direction of Chaplain Ralph C. Deibert. Deibert was the Director of Training, Public Relations Director, and was one of the three new instructors at the chaplain school.[26] However, his administrative skills and public relations abilities were lacking. The Pathe News film showed that the school was not fully functioning. In the film, classes overlapped and barracks rooms held more than their capacity.[27] The Army Chaplaincy had to make adjustments quickly as a result of the public relations disaster.

To rectify this, the Army and Chaplain Arnold created a new film *For God and Country* produced in 1943 by Metro-Goldwyn-Mayer. The film was so accurate in detail the Navy used it as a training film for their chaplains. This film was initially a training film not for chaplains but "to other service personnel in order that they may understand the chaplain's place and duties in the Army."[28] The movie was so influential within the military that the War Department began to show it in movie theaters throughout the United States. When they did this, the Army showed how chaplains were performing and caring for soldiers.

Most movie theaters played military films called "Shorts" because of their 10-20 minutes in length. Unlike the "B" movies, which were over an hour long, *For God and Country*, was 43 minutes long. This new film combined actual chaplain training and professional actors. Chaplain (Lieutenant

Colonel) George J. McMurray, a Southern Baptist minister, was the technical advisor. McMurray entered the chaplaincy in 1920, and provided technical advice. The movie portrayed the chaplaincy as dramatized in the stories of four chaplains. Future United States President Ronald Reagan portrayed the role of the Roman Catholic Priest, Father Michael O'Keefe. Richard Whorf portrayed Rabbi Arnold Miller. Actors Hugh Marlowe and Richard Carlson portrayed the two Protestant Chaplains, Mark Richards, and Tom Manning. Reagan, like many other Hollywood actors, was assigned to the 1st Motion Picture unit of the United States Army Air Force. This unit made approximately 400 training films for the Army.

The publicity of chaplains providing religious support validated the Army chaplaincy to the War Department and the nation. The sensitive and personal nature of religion, especially in time of war, requires constant monitoring and professional oversight to prevent information fratricide.[29] The chaplain's message of goodness and hope, even in the darkest of times, will always be good news for the public and will validate the existence of the chaplaincy for future generations.

Army Air Force Chaplaincy

Chaplain Arnold's vision of the chaplaincy's responsibility to care for soldiers was more significant than any interoffice rivalry. His vision was evident in allowing and supporting the Army Air Force to have its own Division that included their own personnel managing Air Force Chaplains and prescribing the additional training required beyond the Chaplain School.

General George Marshall wanted greater unity of effort and authority for the Army Air Corps, and on June 20, 1941, the Ground Headquarters Air Force and the Office of Chief of Air Corps combined with Major General Henry Harley (Hap) Arnold in command achieved greater unity of effort.

Chaplain Arnold knew General Arnold from their days playing golf together at Fort Leavenworth. One day General Arnold asked him, "Padre, do you think we could be related?"

Arnold replied, "Most Arnolds come from England. My people came from Switzerland."

One day several years later. "Hap" Arnold met Chaplain Arnold in the Pentagon. "Padre." he said, "I'm a sucker for these genealogy fellows, and I hired one in New York to look up my family tree. I find [sic] there were five generations of Arnolds in England and before that they came from Switzerland. How did your people spell their name?"

"Arnoulde."

"That's the way mine did. So maybe we had a grandpappy back there."

Chaplain Arnold smiled and said, "I am Catholic. You are Baptist. Which of us slipped?"[30]

This congenial exchange between General Arnold and Chaplain Arnold shows the importance of building relationships with fellow staff officers.

The importance of highlighting the relationship in the establishment of the Army Air Force is to show the connection between organizational leadership and pastoral ministry in the military. The relationship between leaders and chaplains is not always formal, nor should it be. The interaction between these two men shows that General Arnold was thinking beyond military duties by asking questions about his past, heritage, and family. The representation of the chaplain transcends military life. The chaplain provides an understanding of where we came from, who we are, and where we are going. When a chaplain touches upon these points, even with senior leaders, he provides a sense of balance and purpose to the senior leader. There is no better example of this than the small slips of paper found within the General George C. Marshall papers at the Marshall library. Marshall still kept personal notes from Chaplain Arnold expressing

Arnold's thoughts and prayers for personal family illnesses or difficult situations. Chaplain Arnold's relationship with General Marshall and General Arnold developed into a trusting and enduring relationship that served God and the military.

Chaplain Arnold understood the vision of senior leadership and nested his vision with that of senior leaders around him. One day, Chaplain Arnold said to General Arnold: "It looks as though some day you're going to have your Corps of Chaplains."

General Arnold replied, "That's what I want."[31] With these words, Chaplain Arnold went to work to support General Arnold's vision for the separate Air Force.

In February 1941, Chaplain Arnold wrote to the Adjutant General the following memorandum:

> In order that the higher echelons of the Air Force may have adequate religious ministration supervised by Regular Army chaplains of appropiate [sic] grade, it is recommended that the following named chaplains be relieved from their present duties and assigned to duty as indicated.[32]

With this memorandum, Chaplain Arnold anticipated the future establishment of the Army Air Force. He knew the challenges of the unity of Command and sought to support General Arnold with his own Air Force chaplaincy. Chaplain Arnold saw the future of the Air Force not as a loss, but a way to build upon the foundation of the Army to care for the religious requirements of its service members and families.

Chaplain Arnold personally picked Chaplain Charles Carpenter, a Methodist minister, to lead the Army Air Corps. As early as 1942, Chaplain Arnold envisioned the Air Corps would eventually be a separate service. Arnold told

Carpenter to "establish the chaplaincy for the Air Corps…I know it isn't [in existence yet], but it will be."[33]

Army Chaplain publications reflected the rapid growth, size, and complexity of the Army Air Force Chaplaincy. Before July 23, 1943, Army Regulations, Manuals, and Technical Circulars did not specify chaplain support to Army Air personnel. Change 2 to Army Regulation 60-5, *Chaplains*, July 23, 1943, included the Commanding General, Army Air Forces, as part of the military channel for reports from Army Air Force Chaplains. What this meant was that the Senior Chaplain for the Army Air Force Chaplain now would monitor the reports from his chaplains.[34]

Figure 8. Chaplain Charles Carpenter. Source: Army Chaplain Museum.

The evolution of the monitoring of Army Air Force Chaplain monthly reports shows the importance of assessing and adapting to achieve the objective. Early in the War, all chaplain reports went directly to the Chief of Chaplains' office. Chaplain Carpenter asked the Chief's office to have all Air Corps chaplains forward their reports through the air chain of command to OCCH. The Air Liaison in the Chief's office said that the Deputy Chief of Chaplains, Chaplain George Rixey, said: "You can't do it." Carpenter's office then drafted an order for all Air Corps chaplain reports to go through the Air Chaplain's office. Carpenter then took the order and went to a War Department staff office, told them his situation, and they stamped it "By Order of the Secretary of War." Carpenter recalled what happened next.

> I got a call from the Chief of Chaplains Office that the Chief wanted to see me… I said to him. "I'm glad you called me, because I want to thank you for something you have done." He said. "What's that?" I said, "For publishing that amendment to the 65 series that gives recognition to the Air Corps chain of command." He said, "Oh, I thought you published that." I said "I brought this down a number of times here and I thought you people had gotten it out for me. Well, it's been published anyhow, and it's a thing that's needed." And nothing more was said about.[35]

Changes 2 to Army Regulation 60-5 included this change...

WAR DEPARTMENT
31 July 1943

CHANGES 2
AR 60-5, 19 May 1942, is changed as follows:

4. Duties

h. Reports.-on the last of each month chaplains will render on the prescribed form (W.D., Ch. Form No.3) a report of all duties performed by them during the month. These reports will be typewritten and in duplicate. The original will be forwarded through military channels **(including the Commanding General, Army Air Forces, in the cases of reports rendered by chaplains assigned to the Army Air Forces or units thereof)** to the office of the Chief of Chaplains for his information for file.

BY ORDER OF THE SECRETARY OF WAR:
G.C. MARSHALL, Chief of Staff [36]

Arnold's vision and Carpenter's understanding of that vision and intent facilitated the unity of effort within the Army and Army Air Force Chaplaincy.[37]

The 1944 Technical Manual 16-205, *The Chaplain* shows another example of how OCCH supported the establishment of the Army Air Force. In Figure 12 of the manual, supervisory relationships outlined the Chief of Chaplains to the assignment of chaplain captains. Half of the diagram includes the Army chaplain subordinate functions and units. The other half of the diagram contained the Army Ground Forces Chaplain and the Army Service Forces Chaplain. Although all three component chaplains fall directly under the Chief of Chaplains, only the Army Air Chaplain is supervised directly by the Chief of Chaplains. The other two chaplains have "No Chaplain-Supervision by the Chief of Chaplains." The intent of this diagram shows not that Chaplain Carpenter needed supervision, but that

Chaplain Arnold placed Chaplain Carpenter near and almost equal to himself within the War Department.[38]

The establishment and buildup of the Army Air Force in the United States, Europe, and the Far East, challenged OCCH regarding how to employ chaplains in the newly established arm of the military. During World War I, aviation units fell under the Signal Corps. These units received area coverage by chaplains who were hundreds of miles away. The oversight of the requirement by OCCH for Air Units was essential to the success of making sure that every air unit had a chaplain. The oversight resulted in significant growth of the number of chaplains for air units during World War II. In March 1942, there were 268 chaplains in the Army Air Force (AAF). One year later, there were 1,249 AAF chaplains. By April 1943, all units going into combat areas who had chaplains on their tables of organization had a chaplain. By April 1945, there were 1,861 Chaplains in the AAF.[39] Newly activated Army Air units received authorization for one or more chaplains depending upon the mission of the organization. This new authorization included many training centers and bases in the United States and overseas.

The momentum of the Army Air Force Chaplaincy under Chaplain Arnold changed significantly in June 1945 under Arnold's replacement Chaplain Luther Miller. The original Army-Air Force Agreements discouraged parallel branches in the two services. General Arnold's replacement as Chief of the Air Force, General Carl Spaatz, was ready to meet with Chaplain Miller. Spaatz had already determined to follow the example of the Marines who used Navy chaplains as "borrowed manpower" and allow Army Chaplains to serve in the Air Force. In mid-1948, Spaatz was ready to meet with Chaplain Miller to finalize this relationship. However, when he met with Chaplain Miller at 11 o'clock, Spaatz was disappointed to learn Miller sought to retain oversight and control for the basic training of chaplains, procurement of supplies, and equipment. Spaatz quickly realized there would be no unity of effort and asked Miller to return to his office at 4 o'clock that afternoon. Spaatz's chaplain, Chaplain Charles Carpenter, had a one-page decision paper that included the structure and organization for a separate Air Force Chaplaincy. Spaatz folded the paper to hide Carpenter's signature block and placed it under the glass on his table. When Miller returned in the afternoon, Spaatz showed Miller how he wanted his own Chief of Chaplains.[40] Miller acquiesced and allowed Carpenter to serve as Chief of the Air Force.

When Chaplain Luther Miller met with General Spaatz, he must have forgotten the guidance written in the 1947 edition of *The Chaplain* manual. In the very first paragraph of Section IV, Administration and Supervision of AAF Chaplains, it states, "The Air Chaplain will be appointed by the Commanding General, AAF, from the chaplains on duty with the AAF."[41] Nevertheless, Chaplain Miller still wanted control over the Air Force. Chaplain Carpenter revealed shortly before his death additional information he received about the encounter between General Spaatz and Chaplain Miller. Spaatz went to the meeting willing to use borrowed personnel from the Army. However, when Chaplain Miller said that the new leadership of the Air Force would need a Brigadier General and that Carpenter would return to the Army, and that one of Miller's choices would lead the Air Force Chaplaincy, Spaatz realized that the Air Force Chaplains would always be subservient to the Army.[42]

The chaplaincy is not exempt from the rivalry between the Army and the Air Force. When Chaplain Carpenter received his second star as Chief of Chaplains for the Air Force in 1949, the Air Force Chief of Staff, General Hoyt Vandenberg backdated his rank to 1943, so that he would be the senior chaplain on the newly established Armed Forces Chaplains Board. Carpenter, like Chaplain Arnold, served two terms as Chief of Chaplains from June 11, 1948, to August 14, 1958.[43]

The Army Air Force assignment of chaplains in the numerous support roles highlights the importance of the requirement to have chaplains at these locations. The Army Air Force provides the best example of the importance and dangers of ministry of a chaplain far from the front line. Chaplain (Captain) Keith B. Munro, a Presbyterian chaplain, assigned to the 871st Airborne Engineer Aviation Battalion, a glider engineer unit of the Army Air Force whose mission was to build airfields. On Sunday, August 15, 1943, Chaplain Munro was conducting worship services in the jungle near Tsili Tsili, New Guinea. Overhead Japanese bombers met with Allied fighters, and one Japanese bomber crashed close to the worship service. Chaplain Munro died directing his congregation to safety.[44] His Commander, Colonel Henry G. Woodbury wrote the following to Munro's parents, and to the Bureau of Public Relations.

> He died with God's words on his lips and could he have chosen the time of God's calling, he would have had it thus. He was buried last evening in a grove of coconut palms surrounded by full blooming poinsettias. He gave his life that those he love could live in peace and freedom.[45]

The establishment of the Army Air Force chaplaincy allowed for religious support for a specialized function that resulted in airmen receiving direct religious support all over the world.

Ministry of Presence

The orchestration of chaplain organization began with Chaplain Arnold and OCCH. From there, the organization of the chaplaincy adjusted to ensure soldiers everywhere had a chaplain to provide worship services and care for soldiers. Garrison and regimental chaplains had been the standard since World War I. These organizations covered of a significant amount of geographical area and support units whose function was not direct warfighting. Chaplain Arnold initiated the concept of "ministry of presence." This concept linked soldiers and chaplains at every echelon for every type of organization and unit. However, the doctrinal term "ministry of presence" did not occur until the publication of Field Manual 16-1 *Religious Support Doctrine: The Chaplain and Chaplain Assistant*, November 1989. [46]

Before World War II, chaplains served in regiments and garrisons. During the interwar years and the establishment of the CCC camps, garrison peacetime ministry became a priority for OCCH and the public at large. During the Depression era, husbands and fathers left their homes to work and live in these camps and garrisons in order to provide money to support their loved ones. Active duty and reserve chaplains provided "Garrison Churches" for their religious and spiritual needs. Chaplains and their religious programs protected them from the vices of the world.

These "Garrison Churches" were Jewish, Catholic and Protestant services, but they were a "cooperative effort" to meet the religious needs of the soldiers. If there were enough soldiers of a particular Protestant denomination for a service, a chaplain of that denomination provided the service. Many people understood and accepted the "General Protestant" service.[47] In 1951, a congressional inquiry challenged the establishment of a "General Protestant" service and was soundly defeated.[48] Garrison ministry, and in particular garrison religious services and programs, provided for the free exercise for soldiers and assisted Commanders with the readiness of their soldiers.

These garrison services usually centered on similar missions and units. Chaplains were encouraged to provide for the services of their soldiers and not their own personal needs, as expressed in Circular Letter 218, February 15, 1941.

> Regimental Services—Division and post chaplains are reminded that a chaplain's first duty is to the men of his own regiment. Our Army is a mobile one and the chaplain must be prepared to serve his men under the rigorous conditions of field service. If he does not get acquainted with them and win their confidence while in garrison, he will be handicapped in the field. While an occasional division or post service may be desirable, it is suggested that chaplains of units be encouraged to conduct services regularly for their own organizations.[49]

Chaplain Arnold also publicly rebuked one chaplain without a name who left his post on a Sunday with the sign "There will be no Sunday School or Church on the Post because of the absence of the Chaplain." The severity and implication of this offense struck a nerve with the Chief of Chaplains. In the public rebuke, the Chief highlighted the word "Chaplain" in the post and exclaimed "We wonder!!" if this person truly is a chaplain for those individuals entrusted to his care since someone called by God to care for the people of God would not leave them unattended.[50] Chaplains were to coordinate for coverage and services in their absence because the provision of worship services is the foundation of why the Army has chaplains.

Some chaplain historiography focuses upon the organization and proximity of chaplains at the front line. In this genre, garrison and support role assignments are less desirable and unimportant. Front line ministry is critical and important but not at the expense of neglecting other areas of ministry in the Army. The European Theater of Operations After Action Report highlights the desire of soldiers to have chaplains on the front line. In the report, the Theater G-2 Intelligence Censorship section highlights five cases where soldiers appreciated the chaplain sharing the hardships of combat and being with them on the front line. Two cases of censorship centered around the soldier's dissatisfaction with their chaplains for not being with them on the front line.[51] Similar to Army Air Force Chaplains who wanted to ride on the bombing missions, the ground force chaplains also desired to serve with their flock on the front line. One chaplain who was on the front lines and possibly behind enemy lines wrote this advice to other chaplains in January 1944.

> From this engagement I learned two things: what the infantry soldier sweats out every minute of the day – and that the very front lines is no place for a chaplain. If there had been casualties in the other companies I could not have been present to give the Sacraments to the Catholic boys, nor say a prayer with the non-Catholic boys. I thus learned that the most practical place for a chaplain to be located is at the Battalion Aid Station where he can care for all the wounded.[52]

Each generation of leadership within the chaplaincy must remind chaplains of the importance of providing ministry for all soldiers, regardless of the mission or location. An overview of the doctrinal publications for the organization shows the importance of providing religious support for every organization regardless of their proximity to the front line.

Chaplain Arnold never made front line ministry greater than any other ministry in the chaplaincy. The first official chaplain doctrine manual, Technical Manual 2270-5: *The Chaplain: His Place and Duties* mentions in Section XIII, "The duty of the chaplain lies with the men of his command who are on the fighting line."[53] In this same section, the chaplain's place of duty is at aid stations, hospitals, graves registration, military funerals, pastoral correspondence, censorship, and finally military executions. If this duty was so important, it should have been at the beginning of the manual. Before Section XIII, there were several sections on the chaplain's duty for religious observances, special services, pastoral duties and the chaplain's duty as a staff officer.[54] During World War II, to focus only on chaplains at the front line neglects the organization of the 1,000 plus chaplains in the United States and those in support roles throughout the world. Regardless of whether soldiers were at reception centers, railroad units, transport ships, airfields, army posts, or on the fighting line, the chaplain was always there.

A few chaplains wanted to be on the "front line" and fly on bombing missions. Chaplain Reynolds of the Eighth Air Force corrected this with the following guidance to his chaplains.

> The importance of chaplains being present when the planes take off and when they return from their combat missions is paramount and no one can question the morale value and the spiritual re-enforcement when they take off in the face of possible death or when they return bearing their

> wounded with them. There is undoubtedly a severe mental and nervous tension present at all times in the combat zones on the part of those engaged in combat operations. Good chaplains are invaluable and no others can be used.[55]

This example reflects the importance exhibited within the Army and Air Force chaplaincy to monitor and correct activities taken by subordinate chaplains. Although the motives of subordinate chaplains show their desire to care for soldiers and airmen, supervisory chaplains must exhort and encourage chaplains to understand the greater need in light of limited resources. This principle does not mean chaplains should not take personal risk in order to care for soldiers and airmen, but chaplains must always conduct a cost-benefit analysis and nest their actions within the overall mission of the unit.

The ratio of one chaplain per 1,200 or 1,000 Soldiers varied depending upon the mission and location. For example, chaplain support to the Philippine Scouts and the unassigned recruits at reception centers did not meet the 1,200 authorized personnel requirement.[56] In 1941, as the war raged in Europe and Asia, geographic Corps areas throughout the United States established reception centers and replacement centers. Although the number of personnel assigned to these centers was less than the required 1,200 personnel, the Chief's office recognized the requirement of ministry and assigned several chaplains to these locations. Filling in this gap based upon the requirement and not just the number of personnel occurred with the transport (ship) chaplain, railroad chaplain, and the graves registration unit, as well.

Another example that went contrary to the ratio of one chaplain for every 1,200 soldiers was the Army's Armor Group. The Armor Group performed as a command and control headquarters attached to an Army Corps and was responsible for separate tank battalions within a corps area. Although the Armor Group consisted of less than a total of 110 officers, a warrant officer, and enlisted men, the Armor Group had authorized two chaplains. The function and purpose spanned beyond the 110 soldiers to the independent battalions with the corps area. The Armor Group chaplains coordinated religious support for the independent battalions and pastoral care to the senior leadership within the Armor Group.

An area of presence of increased need of the Army Chaplaincy was for the prison chaplain, prisoner of war camp, and the Japanese internment camps. The technical classification for these camps fell into four groups: "Prisoners of war, men awaiting trial, garrison prisoners who are confined for comparatively short periods, and general prisoners who have been convicted of more serious offenses."[57] Historically, chaplains provided ministry in prisons. However, during World War II, the Army chaplaincy under Chaplain Arnold's leadership expanded the spectrum of religious support into areas not previously considered. Training and ensuring for the free exercise of religion resulted in chaplains exhibiting the highest qualities of human freedom and chaplain identity.

On December 7, 1941, President Roosevelt issued Presidential Proclamation 2525, and then on December 8, 1941, issued Presidential Proclamation 2526, and 2527. These proclamations authorized United States Department of Justice personnel to detain potentially dangerous enemy aliens, people of Japanese, German, and Italian ancestry, some of whom were American citizens living throughout the United States, and place them into internment camps. Chaplains provided religious support to these individuals at the internment camps as part of their unit assignment or as part of their area coverage. Chaplains also validated the credentials of interned clergy and informed the commanding officer of the camp of all religious activity of the interned aliens. The chaplain would coordinate with the head surgeon in the hospital to allow interned clergy to visit the sick.[58] Even in this complex situation, the Army chaplain provided for the free exercise of religion and treated the internees with dignity and respect.

Chaplains selected to serve at prisoner of war (POW) internment camps usually had the required language skills to interact with the prisoners. In addition to serving the religious requirements for Army forces assigned to the camp, chaplains also provided religious support to the prisoners. When chaplains could not do so because of language or religious requirements, local civilian clergy, properly vetted through the military, would provide religious services. Prisoners who were military chaplains or civilian clergy who were in the military, as was the case with many German POWs went through a screening process and performed religious services under the supervision of an Army chaplain.[59] One internment chaplain, Chaplain Hubert Allenby, wrote to the Chaplain school from the Roswell Internment (POW) camp, Roswell, New Mexico, encouraging the school to continue with the training emphasizing care to POWs.[60] The lessons learned from these camps were placed in the 1944 edition of the chaplain manual, "It has been found desirable to march groups of prisoners under guard to adjacent chapels, having them arrive before the usual congregation and occupy the balcony that they might be spared unnecessary humiliation."[61] One Roman Catholic Chaplain who provided religious support to German prisoners wrote to all other Roman Catholic Chaplains the following advice.

1. Always be priests. Avoid discussions of politics and war news; never denounce German rulers.
2. Refuse any assignment that might be interpreted as spying.

3. Be prudent in all you say; you are easily misunderstood. Remember that the background and training of prisoners is different from your own.
4. Be cautious when questioning the men, lest they feel that you are going into matters that are not your business.[62]

This wise counsel by the chaplain shows the delicate balance of retaining the position of a professional Army officer who will not abuse nor seek favor from the enemy, while at the same time providing for their basic religious and spiritual needs. Treating enemy POWs with dignity and respect helped heal the relationship between the warring nations during the post war years.

To assist the internment chaplain with care for POWs, the Chief of Chaplains utilized the American Bible Society, which stated it "will be glad as far as possible to supply Scriptures in such languages as may be required."[63] Providing foreign language material for POWs was necessary for the chaplains providing care for Prisoners of War and the number of first-generation immigrants in the Army.

A somber aspect of ministry to prisoners is that to those under the sentence of death. The United States executed 36 soldiers during World War I. The military did not execute a soldier during the interwar years. Starting on November 6, 1942, the United States Army carried out 157 military executions for multiple murders, rapes, and a single incident of desertion.[64] The chaplain's role and duty to those under the sentence of death increased and was refined to be sensitive and accommodate the prisoner's faith tradition.

The 1926, 1937, and the 1941 chaplain manual included a section titled "Ministration to the doomed-Military Executions." However, the section is included within wartime activities, and the implication is that a military execution would not be of United States service members.

The 1944 manual changed the location of this section and placed it with all other prison chaplain duties. Chaplains were to:

> Secure the help of a clergyman of the faith in which the man was reared, and the religious rites of his church should be provided if he wishes them. The Chaplain will accompany him to the place of execution and do whatever he can to sustain and comfort him to the last.[65]

The RESTRICTED and first War Department Pamphlet for the *Procedure For Military Executions* published in 1944 directed:

> In all executions, a chaplain of the prisoner's choice will be provided if practicable. If no chaplain of the prisoner's choice or of his particular denomination is available, the officer charged with the execution of the sentence will take all reasonable measures to provide a civilian clergyman from that denomination.[66]

The 1947 edition added additional instruction (emphasis mine):

> In all executions, a chaplain of the prisoner's choice will be provided if practicable. If no chaplain of the prisoner's choice or of his **particular faith and/or race** is available, the officer charged with the execution of the sentence will take all reasonable measures to provide a civilian clergyman of that **faith and/or race**.[67]

The 1947 edition shows the great length the military took to accommodate the prisoner's religious beliefs, as well as racial sensitivities. Also, the removal of the Christian term "denomination" to the more generic "faith" reveals the existence of non-Christian prisoners executed for war crimes. The addition of a section for "interpreters" shows that many of the post-war executions were for enemy war criminals.[68]

The duration and geographical scope of World War II challenged previous concepts of how and where chaplains provide religious support. Continual assessment and adapting while always ensuring religious liberty regardless of the organization, friendly or enemy soldiers, the Army chaplaincy was always there.

The restructuring of OCCH included competent leaders who continually assessed and adapted that resulted in enhanced capabilities of providing religious support. The Army and Air Force Chaplains are part of that legacy regardless of the location or the type of ministry they provided during the war. Telling the story of the chaplaincy's adaptation and transition to care for the men and women in the military is an example for current planners and will ensure the existence and validity of the chaplain for future generations.

Notes

[1] U.S. Department of the Army, Army Regulation 5-22, *The Army Force Modernization Proponent System*, (Washington, DC: U.S. Department of the Army, October 28, 2015), 13.

[2] Headquarters Army Service Forces, Circular Letter 5, Office of the Chief of Chaplains, March 23, 1926.

[3] Augustus A. Goodyear, "Organization and Administration," *The Army Chaplain: Official Publication of the Chaplains' Association of the Army of the United States*, Vol VIII, No 3, (January 1938): 103.

[4] Interview with A. S. Goodyear November 21, 1958, quoted in Jorgensen, *The Service of Chaplains to Army Air Units, 1917-46*, 90.

[5] David B. Jorgensen, *Air Force Chaplains, 1947-1960*, (Volume II, Office, Chief of Air Force, 1961), 379.

[6] Circular Letter 242, Office of the Chief Chaplains. January 2, 1942.

[7] Robert Nay, "The Operational, Social, and Religious Influence Upon the Army Chaplain Field Manual, 1926-1952" (Master's thesis, Command and General Staff College, 2008), 51.

[8] Roy J Honeywell, *Chaplains of the United States Army,* (Washington, DC: Office, Chief of Chaplains, 1958).

[9] Ibid., 242.

[10] Daniel B. Jorgensen, *The Service of Chaplains to Army Air Units, 1917-46,* (Wash, DC: Office, Chief of Air Force Chaplains, 1961), 48.

[11] Robert L. Gushwa, *The Best And Worst of Times: The United States Army Chaplaincy 1920-1945*, (Volume IV, Office of the Chief of Chaplains, Department of the Army, 1977), 99.

[12] Form No. 64, A.G.O, Ed. May 31-17.

[13] W.D. Ch. Form No. 3, September 1. 1942 and W.D. Ch. Form No. 3, March 12. 1943, and W.D. Ch. Form No. 3, March 1944

[14] Army Service Forces, *The Chaplain Serves: Chaplain Activities, 1943*, (Washington D.C. Office of the Chief of Chaplains 1 March 1944), 58.

[15] The Mobilization Regulations 1-10, *Morale*, March 5, 1943 places Religious instruction, service and ministrations as one of twelve factors affecting morale.

[16] *Religious and Spiritual Morale Factors,* The Chaplain School, Carlisle Barracks, Pennsylvania, December 1, 1948. and *ST 16-155 The Chaplain and Military Morale* (The Chaplain School, Carlisle Barracks, Pennsylvania, August 1950).

[17] *ST 16-152 Chaplain Supervision: General Considerations* (The Chaplain School, Carlisle Barracks, Pennsylvania, June 1, 1950), 5.

[18] Circular Letter 202, Office of the Chief of Chaplains, April 2, 1940.

[19] Army Service Forces, Office of the Chief of Chaplains, *The Chaplain Serves: Chaplain Activities, 1943*, (Washington D.C. Office of the Chief of Chaplains 1 March 1944), 22-23.

[20] Circular Letter 269, Office of the Chief of Chaplains, March 1, 1943.

[21] Circular Letter 230, Office of the Chief of Chaplains, July 15, 1941.

[22] Circular Letter 259, Office of the Chief of Chaplains, September 15, 1942.

[23] *The Chaplain Serves: Chaplain Activities, 1943*, 9.

[24] Ibid., 23-25.

[25] Circular Letter 268, Office of the Chief of Chaplains, February 15, 1943.

[26] Army Service Forces, United States Army Chaplain School, *History of USACHS: France 1918-Havard University* (Unpublished document prior to the establishment of USACHCS, original onion skin paper copy with Chaplain (Colonel) Robert Nay), 23.

[27] Gushwa, 109.

[28] *The Chaplain Serves: Chaplain Activities, 1943*, 53.

[29] Information fratricide, which is defined as adverse effects on the information environment resulting from a failure to effectively synchronize the employment of multiple information-related capabilities which may impede the conduct of friendly operations or adversely affect friendly forces. *Field Manual 3-13, Information Operations*, (Department of the Army, December 2016), 1-33.

[30] Jorgensen, *The Service of Chaplains to Army Air Units, 1917-46*, 87.

[31] Ibid., 99.

[32] Ibid., 97.

[33] John E. Groh, "Summary History Of The Air Force Chaplaincy" (PB 16-90-1, Military Chaplains Review, Winter 1990), 72.

[34] U.S. War Department, Army Regulation 60-5, *Chaplains* Change 2, July 31, 1943, (War Department, May 19, 1942).

[35] *Summary History Of The Air Force Chaplaincy*, 71.

[36] U.S. War Department, Army Regulation 60-5 Change 2, *Chaplains: General Provisions*, (War Department Washington D.C.) July 31, 1943.

[37] It must be noted that Chaplain Arnold did not recommend his deputy, Chaplain Rixey to replace him as the Chief of Chaplains. Chaplain Luther Miller became the next Chief of Chaplains and Chaplain William Cleary as the next deputy. Arnold took Rixey with him as his deputy for the Inspector General. Rixey kept his rank as Brigadier

General until his position in the IG was abolished. Rixey then reverted back to Colonel April 1, 1946 and served as the 2nd Army Commander and then as the XXIV Corps Chaplain in Korea. Rixey retired March 31, 1948 and advanced to Brigadier General in retirement on June 29, 1948.

[38] U.S. War Department, Technical Manual 16-205, *The Chaplain,* (Washington, DC: War Department, 5 July 1944), 75.

[39] Jorgensen, *The Service of Chaplains to Army Air Units, 1917-46*, 101.

[40] Jorgensen, *Air Force Chaplain, 1947-1960*, 6-7.

[41] U.S. War Department, Technical Manual 16-205, *The Chaplain,* (War Department, June, 1947), 4.

[42] Interview of Chaplain (Major General, Retired) Charles I. Carpenter by Chaplain (Colonel) James M. Thurman, March 25, 1987, quoted in *PB 16-90-1 Military Chaplains Review* Winter 1990 "Summary History Of The Air Force Chaplaincy" John E. Groh, 75.

[43] *Summary History Of The Air Force Chaplaincy*, 71.

[44] Army Forces Pacific, *History of Chaplains' Activities in the Pacific*, (Chaplain Section, Ground Headquarters, Army Forces Pacific, 1946), 205-206.

[45] Berkeley Daily Gazette November 6, 1943, page 1.

[46] The doctrinal term "Ministry of Presence" did not occur until Field Manual 16-1 *Religious Support Doctrine: The Chaplain and Chaplain Assistant*, November 1989, page 5-20.

[47] Jorgensen, *The Service of Chaplains to Army Air Units, 1917-46*, 70.

[48] *Summary History Of The Air Force*, 83.

[49] Circular Letter 218, Office of the Chief of Chaplains, February 15, 1941.

[50] Circular Letter 232, Office of the Chief of Chaplains, August 15, 1941.

[51] U.S. Forces, European Theater. *Report on the Army Chaplain in the European Theater*, Appendix 16.

[52] Military Ordinariate Circular Letter 35 (462 Madison Avenue, NY, NY, January 31, 1944), 2.

[53] U.S. War Department, Technical Manual 2270-5, *The Chaplain: His place and duties*, (Government Printing Office, 1926), 51.

[54] Technical Manual 2270-5: *The Chaplain: His place and duties*, 1926, 35-45.

[55] Jorgensen, *The Service of Chaplains to Army Air Units, 1917-46*, 132.

[56] U.S. War Department, Army Regulation 60-5, *Chaplains*, (Washington, DC: War Department, December 20, 1937), para 1.

[57] U.S. War Department, *The Chaplain,* Technical Manual 16-205, (Washington, DC: War Department, 5 July 1944), 48.

[58] Memorandum from Office of the Chaplain, *The Prisoner of War and Internment Camp Chaplain*, Headquarters, Stringtown Internment Camp, Stringtown, Oklahoma, March 15, 1943, to Brigadier General B.M. Bryan, Director, Aliens Division.

[59] *The Chaplain Serves: Chaplain Activities, 1943*, 56-57.

[60] *History of USACHS: France 1918-Havard University*, 52.

[61] Technical Manual 16-205, *The Chaplain,* 36.

[62] Military Ordinariate Circular Letter 40, (462 Madison Avenue NY, NY, October 1, 1944), 4.

[63] Circular Letter 262, November 2, 1942.

[64] https://www.census.gov/prod/2001pubs/statab/sec05.pdf (accessed May 29, 2018).

[65] Technical Manual 16-205, *The Chaplain,* 49.

[66] U.S. War Department, War Department Pamphlet 27-4 RESTRICTED, *Procedure For Military Executions*, (War Department, June 12, 1944), 2.

[67] U.S. War Department, War Department Pamphlet 27-4 RESTRICTED, *Procedure For Military Executions*, (War Department, December 1947), 3.

[68] Ibid., 3.

Chapter 4

Training

Training is integral to the success of any military, and the training of chaplains is an important part of that. Army Chaplains are trained professionals providing a unique capability for an effective and disciplined force. The Army defines training as "The instruction of personnel to increase their capacity to perform specific military functions and associated individual and collective tasks."[1] A suggested religious support definition is, "The instruction of professional military, religious leaders and professional religious advisors (56A and 56M) to increase their capacity to provide and perform with other religious and faith groups."

The Army Chaplaincy developed three efforts to meet the training capability gap. First, it instituted Army correspondence courses in a time of limited resources. Second, Chaplain Arnold started the process for the reactivation of the Chaplain School. Third, senior leaders within the Army Chaplaincy understood the complexity and importance of training civilian clergy to perform as chaplains. These priorities show the importance Chaplain Arnold took to provide training for chaplains during World War II.

Army Correspondence Courses and Supervisory Chaplains Supporting Training

The chaplain school at Fort Leavenworth, Kansas, provided its last resident class in 1928. Low enrollment and lack of funds forced the closure of the school for students. However, the chaplain school remained open and consisted of two personnel who were "staff and faculty in addition to post duties."[2] By 1940, the chaplain school had two hundred pounds of records, a library of fifteen books, ten framed pictures of past classes, and a fund of $102.92, and no students.[3] The Army chaplaincy used correspondence courses during the interwar years to meet the gap caused by the closure of the school. (See Appendix 4 for the list of correspondence courses provided during the interwar years).

During the interwar period, chaplains had to complete the prescribed correspondence courses and attend summer camps within their first five years in the chaplaincy, in order for consideration for retention and promotion. Managing this by OCCH was essential to meet the 1925 mobilization plan that called for 1,870 Regular Army and Reserve Chaplains. During the interwar years, the Army Chaplain School never met this requirement for trained personnel.[4] The major reasons for not meeting the requirement was the lack of chaplains at the Chaplain School and poor supervision by chaplains.

The 1941 edition of Technical Manual 16-205, *The Chaplain*, outlined the purpose of the Chaplain School.

> The mission of this school is to give chaplains such special training as may be calculated to fit them to minister in a comprehensive, liberal, and efficient way to the moral and religious needs of the military service both in peace and war. During periods when no sessions are held, its mission is the preparation and revision of correspondence course material for use by the War Department extension schools.[5]

Because the Chaplain School did not fulfill the requirement to train chaplains, in 1940 Chaplain Arnold directed that OCCH oversee the correspondence course program due to its lack of completion by chaplains.[6]

Even with Japanese aggression in Asia and German expansion in Europe, chaplains still did not enroll nor complete the required extension courses.[7] One contributing factor for chaplains not completing the courses was the lack of chaplain supervisors providing mentoring, oversight, and accountability for their training.

There is only a brief paragraph in the manuals exhorting chaplains to supervise subordinate chaplains. "He is charged with the duty of coordinating and supervising all religious work in the Army and from time to time making recommendations for such action as he deems advisable to promote the moral and spiritual welfare and contentment of the Army."[8] The manuals do not provide guidance or techniques for coaching or teaching subordinate chaplains with their work. However, in the 1944 manual, a section of nine pages is devoted to the roles and responsibilities of supervisory chaplains at every level of war. Shortly after World War II the one area that received additional attention and increased publication of instructional documents was the training of supervisory chaplains.

The first three chaplain manuals, starting in 1926 through 1941 stated, "Chaplains will be expected to take courses of study in the troop schools for officers, along with officers of line and staff." These courses augmented correspondence courses. Both forms of education were military in substance, with minimal instruction on "Practical Duties" for chaplains. However, part of the reason for the failure of chaplains to complete the correspondence course is the ambiguous wording found in the manual under "Supervision."

> For the manner in which he performs his military duties the chaplain is accountable to his commanding officer. In ecclesiastical matters the chaplain conforms to the requirements and practice of the particular denomination to which

he belongs. In matters touching upon methods of work and professional policy which do not involve church doctrine, supervisory authority is vested in the Chief of Chaplains. He is charged with the duty of coordinating and supervising all religious work in the Army and from time to time making recommendations for such action as he deems advisable to promote the moral and spiritual welfare and contentment of the Army.[9]

It appears from the wording that only the Chief of Chaplains can make recommendations about the methods of providing religious support, including professional training for subordinate chaplains. In May 1942, OCCH published *Helpful Hints and Other Information.*[10] In this document, supervisory chaplains was further weakened and reminded that their "rank does not include command, that they must not give orders, to other chaplains or anyone else, that their sole responsibility as supervising chaplains is to help the regimental or unit chaplains." Chaplains were warned, "against becoming too military or too secular in their attitude toward each other or toward their work."[11]

This challenge and tension of transforming civilian clergy into military chaplains reached its zenith at the Army Chaplain School during World War II. The Commandant at the Chaplain School wrote that the students lacked the military discipline required for the Army Chaplaincy.[12] At the start of World War II, the first revision of Army Regulation 60-5, *The Chaplain* addresses supervision. The regional Corps Chaplains were to:

...advise and assist them and coordinate their several activities as chaplains; they will provide

> candidates for appointment or promotion with such forms and information as may be necessary; they will make recommendations concerning assignment of Reserve chaplains to active duty in the corps area; they will endeavor to procure a sufficient number to meet the requirements of the corps area and will act as director in the chaplain's branch in the Army Extension School.[13]

In 1943, the chaplain school developed an advance course to train chaplains for supervisory positions.[14] The Army Chaplaincy learned the importance of supervisory chaplains reinforcing the institutional training from the Chaplain School.

The Army Chaplain School

When Chaplain Arnold became the Chief of Chaplains in December 1937, he directed Chaplain Arthur C. Piepkorn to conduct an in-depth study on the Chaplain School. His reasoning was the need to rapidly train thousands of new clergy and transform them into Army chaplains. Chaplain Arnold considered using the United Kingdom Chaplain School in England but feared that chaplains would deploy to the front before the completion of their training. Chaplain Arnold discovered that during World War I, the tendency at that time was to send the chaplains to the front before the completion of their training.[15]

The following information is from an unpublished document from that in-depth study. On August 26, 1940, OCCH proposed a Table of Organization for a reactivated Chaplain's School at Fort Leavenworth as part of the Office of the Assistant Chief of staff G-1. This proposal plus several supporting memorandums resulted in no activation of a chaplain school. On March 28, 1941, the Assistant Chief of Staff G-3 approved the proposed changes from OCCH, with only minor amendments. On April 3, the Secretary of War directed the publication of the amended changes. Publication took place in Section II, Circular No. 70, War Department, dated April 15, 1941. Additions were made "to provide instruction as to administration, coordination, and supervision of chaplains' activities within divisions and higher tactical and territorial units" by the creation of a "department of staff administration, coordination, and supervision" with appropriate subjects in the curriculum. Also, the beginning dates and the length of all courses were to be "as prescribed in orders convening each session." Meanwhile, at the War Department, additional memorandums circulated with no formal activation of the chaplain school. Then on December 9, 1941, two days after the attack on Pearl Harbor, Chaplain Arnold again requested, and the War Department approved, the reactivation of the Chaplain School.[16]

In the same request for the activation of the Chaplain School, Chaplain Arnold wanted the school near Washington D.C., in the interest of adequate supervision by OCCH. Chaplain Arnold did this because he learned from the lessons of World War I. During World War I, many chaplains arrived in Europe with no military training. Therefore, the Allied Expeditionary Force established a chaplain school in France. The courses at the school in France was "informal and without method."[17] The reality of the need for chaplains at the front meant that chaplains received little or no training before arriving at their assigned unit.[18] Chaplain Arnold wanted to ensure chaplains received proper training before their deployment overseas.

Another interesting aspect of the training chaplains that could also apply under the domain of policy concerned the deployment of Chaplains overseas. Although Chaplains and Medical Personnel could deploy immediately overseas after completion of their military training, Chaplain Arnold required chaplains to remain in the United States for four to six months, receiving further training and oversight.[19] Although this caused some tension, as mentioned in the European Theater of Operations After Action Report, the leaders in Europe did not know the background of the deficiencies in World War I.[20]

On January 12, 1942, Chaplain Arnold selected Chaplain (Lieutenant Colonel) William D. Cleary to lead the reactivation of the Chaplain School at Fort Benjamin Harrison, Indiana. Chaplain Cleary took with him Chaplain Herman Page as his secretary and four other chaplains as instructors.

The Army Chaplain School quickly exceeded the competing resources by other Army services at Fort Benjamin Harrison. Chaplain Cleary and Chaplain Page conducted a course of action development and comparison of possible locations throughout the United States and found two civilian schools willing and available to host the Army Chaplain School. The two schools were Harvard University in Boston, Massachusetts and Duke University at Chapel Hill, NC. A personal inspection of both locations by Cleary and Page determined Harvard as the ideal choice for the military. The primary reason for choosing Harvard over Duke centered on the segregation laws in North Carolina.[21] Neither location had an officer mess for the students to dine in. Students would have to use the school cafeteria and local establishments. The segregation laws in North Carolina meant that African American students would dine separately from other students. Restricting liberty and freedom was contrary to Chaplain Arnold's and Chaplain Cleary's vision for two primary reasons.

First, the Army Chaplain School was the first military school to integrate their training. Students trained and lived together regardless of their race. Second, the administrators at the Chaplain School would place students who were senior in rank and more experience yet who had not been to the Chaplain School, with chaplains with little experience with the military. The intent was for the senior chaplains to mentor the less experienced chaplains. However, in February 1943, Chaplain Cleary assessed and developed a new plan that placed students of different denominations together in the same living quarters. The segregation laws in North Carolina would not allow Chaplain Cleary's vision for the school. The goal was to break down barriers between religious groups in order to have a working relationship.[22] The 1944 Technical Manual 16-205, *The Chaplain* explained this concept.

> As the Army consists of men reared in all forms of religious faith and accustomed to a great variety of practices, chaplains must gain a sympathetic acquaintance with the beliefs and usages of those with whom they will have most frequent contact. This is especially true of other chaplains. A fuller understanding of the ideals and purposes of sincere people of other faiths invariably leads to an enhanced respect for them and a more intelligent basis of cooperation.[23]

Establishing trust among the chaplains is necessary before any supervisory and mentorship can happen.

Harvard also had historical religious significance because of the religious freedom views of its founders. During the American Revolution, Continental troops stayed on the grounds of Harvard. George Washington spent a significant amount of time at Harvard, even staying at the home of the College President. The President of Harvard at the time, Samuel Langdon, was himself a former chaplain in the French and Indian War of King George's Army. During the Revolutionary War, on the eve of the Battle of Bunker Hill, Langdon went to the front and prayed for soldiers.[24] Harvard's other leadership also had a helpful attitude toward the chaplaincy. For the Army, Honeywell captures in detail the cost of training chaplains at the university.

> The university provided space for offices and classrooms at an annual charge of 30 cents a square foot. Dormitory space cost $3.56 a week for the first 175 students and 75 cents for each man above that number. Other important accommodations were furnished without charge. The 12 sessions beginning in October 1942 had an average attendance of 375 and cost for these facilities about $10.50 a student. The adjutant computed that it cost 4.38 cents to instruct each of the 440 students of the December class for 1 hour.[25]

During the dark and gloomy days during 1942, 1943, and into 1944, the Chaplain School at Harvard University continued to faithfully train Army Chaplains. The historical significance of the location at Harvard that included religious liberty and equality contributed to the enduring mission and identity of the chaplaincy to provide religious support to all service members. Wartime production of military posts meant the chaplain school would move again, this time to an Army post.

The Army Chaplain School had four subsequent locations during the War. The first class at Fort Benjamin Harrison started on February 2, 1942, with 75 chaplains. Each class ran 28 days, with 200 hours of military instruction that included the required courses from the War Department and additional classes on the military chaplaincy. The school moved to Harvard University in the summer of 1942, and again in the summer of 1944 to Fort Devens, Massachusetts. This location allowed chaplains to exercise amphibious landings with the rest of the Army. Then in July 1945, the War Department directed the location of all the Service of Supply schools located at Fort Oglethorpe, Georgia. Even though the school was in the South, the school remained integrated.[26] However, tension remained between the school and the local population. In January 1947, the school moved to Carlisle Barracks, Pennsylvania; Jim Crow laws were not in effect there at that time. Then exactly two years after the school moved to Carlisle Barracks, President Harry S. Truman issued Executive Order 9981 that abolished discrimination on the bases of race, color, religious, or national origin in the United States Armed Forces.[27] The Army Chaplaincy as an institution for religious freedom was ahead of its time with the integration of training for chaplains regardless of their race.

Chaplain Carpenter of the Army Air Force assessed and determined additional training for Army Air Force chaplains and chaplain assistants. In 1944, at the completion of the Army chaplain basic course, all Army Air Force Chaplains attended a two-week "Air Chaplain Transition Course" at the San Antonio Aviation Cadet Center, Kelly Army Air Field, San Antonio, Texas.[28] Chaplain Carpenter required the two-week course because of the unique challenges chaplains would face. In Honeywell's classic work, he writes, "Different combat arms have distinctive psychological problems which affect the chaplain's work. This is especially true of the Army Air Force, where fighting conditions sometimes tend toward recklessness or a degree of fatalism."[29] The Air Chaplain Transition Course contributed to a strong Chaplain Air Force identity.

The Army Chaplain School could not immediately train the high number of clergy who volunteered after the declaration of war. To retain the interest of those who wanted to join, candidates went through a medical examination, received military clothing and in-processed through the interwar era regional corps area that was throughout the United States, while waiting for a slot to attend the school. These corps areas became part of the Army Service Forces. Several states were grouped together to form the Service Command. For example, First Service Command included Connecticut, New Hampshire, Maine, Massachusetts, Rhode Island, and Vermont. Regular Army Chaplains would oversee the training of the new chaplains who resided in this geographic area. Reserve or National Guard Chaplains would conduct the training only when Regular Army Chaplains were not available. The instruction material would consist of the Army Extension Course material and familiarization with Technical Manual 16-205, *The Chaplain.*[30] In addition to their training, new chaplains also performed chaplain duties for soldiers who were also waiting to enter basic training. Throughout this process, a senior experienced chaplain provided advice, guidance, and pastoral care to subordinate chaplains.

Some chaplains never went to the Chaplain School and only received the training at these Corps Areas before their deployment overseas. Those chaplains who did not go to the chaplain school either had prior military service or had connections within the bureaucracy. One of those chaplains was Clark V. Poling. Poling was one of the four chaplains on the ill-fated U.S.A.T. Dorchester.

Training Civilian Clergy for The Army Chaplaincy

Chaplain Arnold and the leadership at the Chaplain School recognized two key factors that assisted with the transition from civilian clergy to the military chaplaincy. First, new chaplains needed to understand military discipline. Second, the instructors at the chaplain school had to model military discipline and character.

In 1944, Chaplain (Colonel) William D. Cleary, Commandant of the Army Chaplain School, wrote in a book about the challenges of training chaplains at the chaplain school:

> Duties of the chaplain require many qualities and abilities which are completely foreign to the daily activities of the average priest, minister, or rabbi in civil [sic] life. It is part of the Chaplain School to help in the difficult adjustment and transition of men of all faiths, [African American, ed.] and white, from all areas of the country, to the most honorable status of chaplain in the United States Army. One of the greatest challenges in training students who will be chaplains is the almost complete lack of formal discipline to which many of the students were subjected in the course of their civil life...His most difficult adjustment is to the rigid requirements of discipline in the Army, for, being a soldier, albeit without arms, he is subject to the same rules of discipline which govern our fighting men.[31]

To help chaplains learn military discipline, Chaplain Cleary went to great lengths to provide a conducive atmosphere at the school. In doing this as the leader at the chaplain school, Cleary echoed what his supervisor, Chaplain Arnold, wrote earlier when he became Chief of Chaplains.

> Wearing the uniform, with all that it signifies and imposes, is not a hindrance to spiritual life. Rather,

> it is a definite help. Discipline makes a man an efficient soldier in the Kingdom of God as well as in the kingdom of man.[32]

Chaplain Cleary, his staff, and the instructors at the Chaplain School modeled how professional clergy answered their call and performed as chaplains in a disciplined environment.

Insight into the profession of civilian clergy will provide answers to this challenge. Civilian clergy have disciplined behavior. Sermon and worship preparation requires attention to detail. However, the only event they are required to be at is their scheduled worship service and an occasional crisis or meeting. The disciplined requirement to attend events and participate in functions that are not spiritual in nature such as marching formations or map reading exercises presents a challenge for civilian clergy. These non-spiritual duties may seem foreign, and of questionable relevance and significance. However, both military and civilian clergy professions require personal self-development through reading. Civilian education for clergy requires knowledge from known source material. In the military, the soldiers and especially officers and senior non-commissioned officers must know military publications. Students at the Chaplain School received the current doctrinal publication *The Chaplain.* The author of the 1944 edition was the Reserve Chaplain Roy Honeywell. As mentioned earlier, Honeywell was in the Plans and Operations section of OCCH, and in his own words, "with long experience in education in the belief that he would understand the mind of the citizen soldier."[33] During World War II, the instructors at the chaplain school modeled this transition from civilian to military life by their discipline, professionalism, and knowledge of Army publications in the performance of their duties. (See Appendix 5 for the list of classes taught at the Chaplain School)

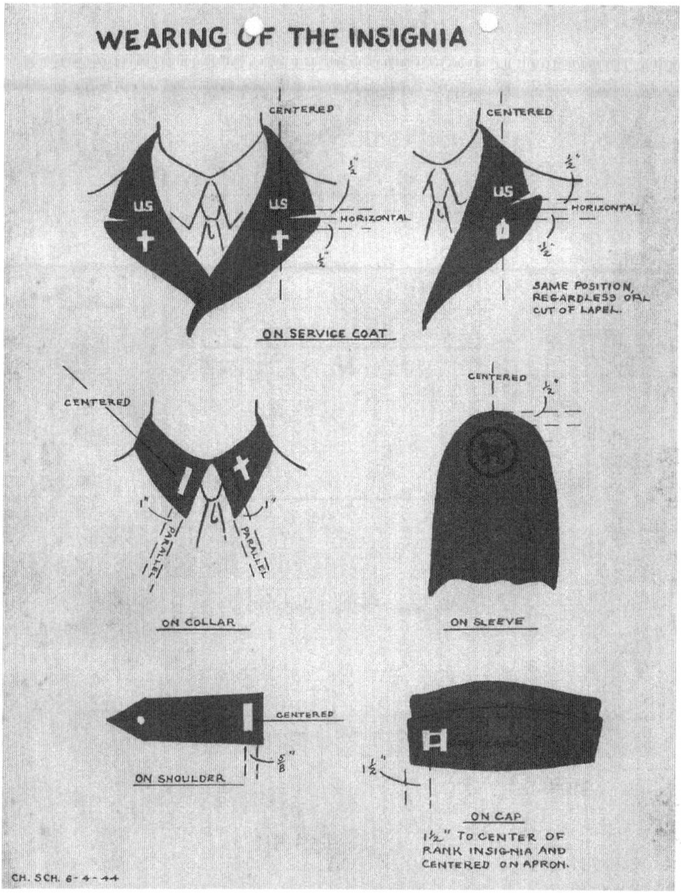

Figure 9. Chaplain School Instruction Handout for Wearing of the Insignia
Source: The author's personal collection.

On February 23, 1943, Chaplain Cleary published a *Manual For Instructors*. In just over 150 pages, the manual sought to "assure coordination and uniformity of instruction under the supervision of the Department of Training at The Chaplain School." The manual presents several methods of instruction and hindrances to learning. Chaplain Cleary exhorts the instructors to "Never decide that the student is stupid" and "never use ridicule…. [or] talk down to a class."[34] Students at the chaplain school already completed a professional degree in their field. As the final report from the Chaplain Activities in the Pacific points out, "The average chaplain here is a fellow with a high school and college education, who has had three years in a seminary and three more as an active preacher, priest or rabbi of a regular church. He has passed stiff verbal and written examinations."[35] Chaplain Cleary also pointed-out that when instructors treat the students disrespectfully or ridicule them, "the students are helpless to retort, their resentment is aroused. When an individual is resentful, his mind is closed to the acceptance of instruction."[36] The manual also included an example of a student's evaluation of an instructor.[37] There are 24 different elements from neatness, posture, dignity, and knowledge, to discipline and management. (See Appendix 6 for the student evaluation) In many ways, instructors at the chaplain school performed as supervisory chaplains. Both must instruct and monitor. The instructors must master the very things expected from the students. Chaplain Arnold wanted all supervisory chaplains to include those at the Chaplain School with these character traits.

Chaplain Milton Beebe, a Methodist chaplain who started his military career in 1916, epitomized the character traits of an Army chaplain. In November 1919, one year after the end of World War I, Chaplain John Axton who went on to become the first Chief of Chaplains in 1920 wrote that, "He is a beaver for work, cheerful, splendid, the best characterization I could give him."[38] In 1939, Beebe captured the importance of military training at the annual convention for Army Chaplains. Although he was referencing the Extension Courses, the same applies to the training at the chaplain school. He begins with, "Every chaplain is presumed to be properly trained." He then goes further and exhorts the students.

> The courses available to chaplains are varied and very informing. They begin with the basic organization of the army, without which every chaplain is a misfit, and carry through a great deal of training that would be helpful to all officers of the army and which is essential to every chaplain who wants to do more than just be carried on the rolls and draw his pay. Those courses are designed to give one "the feel" of the military and to enable him to adjust himself to work with men in a condition that is entirely new and somewhat abnormal. The traditions, customs and practices that are peculiar to the military service are definitely stated, together with the reasons for their existence and men are enabled by consistent study to so absorb the spirit of the Army as to talk intelligently with military men and to share in the bond of fellowship that always exists among comrades in arms who wear the same uniform.[39]

Reading Beebe recommendations reveals that a chaplain must make an effort to learn. A chaplain must not rest upon his previous achievements or education. Beebe himself always made an effort to serve and to learn. Beebe was fifty-five years old in 1941, and had served his nation for twenty-five years. Instead of retiring, he became the Senior Chaplain for the Mediterranean Theater of Operations and oversaw the invasion of North Africa and Italy.

Discipline became the hallmark for those chaplains selected from the Army of the United States for a Regular Army Commission. Chaplain (Major General) Luther Miller's congratulatory letter to appointees highlights the importance of discipline for chaplains.

> Military life is a life of discipline, and the essential virtues of courage, loyalty, obedience, devotion and self-sacrifice are religious virtues. We can only bring God's blessing and the fruits of our ministrations to men and officers by disciplining ourselves spiritually. "Act on the Word, instead of merely listening to it and deluding yourselves" is an admonition that is applicable to all religions and mankind.[40]

As with all branches of the military, not everyone has the discipline to remain in the Army profession. Training and leadership provides the opportunity for anyone who has the desire to be an Army professional.

During the War, there were 8,302 chaplains enrolled in thirty-five sessions at the Chaplain School. Of those enrolled, 8,183 graduated. Only 119 did not complete the course. Fifty-five students left before the end of the first session. Fifty more students failed for academic reasons. The remaining sixty-four students failed because of sickness or other emergency situations.[41] During World War I, faculty at the Chaplain School could recommend to the Army dismissal of any student for any reason. During World War II, the need for chaplains was so great that although many students failed academically, they still served as Army chaplains. A study of the performance of those who failed the Chaplain School showed their performance as chaplains was very low. However, three chaplains failed the Chaplain School but went on to receive high ratings by their commanders.[42]

The United States Constitution allows for a small standing Army, and the mobilization of the citizen-soldier is one way the nation protects and defends itself in times of War. God calls civilian religious and spiritual leaders to serve their people wherever they may be. The requirement to transform a peacetime Army chaplaincy and civilian clergy into a trained and professional Army Chaplaincy required persistence with the bureaucracy, learning from past mobilizations and ensuring only certified civilian clergy perform as professional Army chaplains. During World War II, civilian religious and spiritual leaders felt called by God to serve as Army Chaplains. They volunteered, and the Army trained them to perform their sacred calling in the harshest of circumstances. Chaplains in the Chaplain School exercised their calling as chaplains by training, caring, and ministering to other chaplains. The trainers and training at the chaplain school is the soil for a rich chaplain identity.

Notes

[1] U.S. Department of the Army, Army Regulation 5-22, *The Army Force Modernization Proponent System*, (Washington, DC: U.S. Department of the Army, October 28, 2015), 14.

[2] The Army Chaplain, Volume VII, Number 3, R. Earl Boyd ed, *Annual Report, Chief of Chaplains* (Chaplains' Association of the Army of the United States, January 1937), 12.

[3] Headquarters Army Service Forces, United States Army Chaplain School, *History of USACHS: France 1918-Havard University* (Unpublished document prior to the establishment of USACHCS, original paper copy with Chaplain (Colonel) Robert Nay).

[4] Daniel B. Jorgensen, *The Service of Chaplains to Army Air Units, 1917-46,* (Wash, DC: Office, Chief of Air Force Chaplains, 1961), 52.

[5] U.S. War Department, Technical Manual 16-205, *The Chaplain,* (War Department, April 21, 1941), 6.

[6] *History of USACHS.*

[7] Headquarters Army Service Forces, Circular Letter 198, (Office of the Chief of Chaplains, August 12, 1939).

[8] Technical Manual 16-205, *The Chaplain*, 1941, 11.

[9] Ibid.

[10] Headquarters Army Service Forces, Office of the Chief of Chaplains, *Helpful Hints and Other Information*, (Office of the Chief of Chaplains, May 1942) Original lithoprinted with Chaplain (Colonel) Robert Nay.

[11] Ibid.

[12] William D. Cleary, "How Army Chaplains Are Trained," in *Faith Of Our Fighters,* ed. Ellwood C. Nance (Bethany Press, 1944), 106-107.

[13] U.S. War Department, Army Regulation 60-5, *Chaplains*, (Government Printing Office, May 19, 1942), 4.

[14] Robert L. Gushwa, *The Best And Worst of Times: The United States Army Chaplaincy 1920-1945*, (Volume IV, Office of the Chief of Chaplains, Department of the Army, 1977), 110.

[15] *History of USACHS: France 1918-Havard University*.

[16] Ibid.

[17] Letter from Rev James F. Houlihan, Major (Ret) to Chaplain Arthur C. Piepkorn, Chaplain School, 24 December 1943. Original with Chaplain (Colonel) Robert Nay.

[18] Letter from General Commission Army and Navy Chaplains to Chaplain Arthur C. Piepkorn, Chaplain School, 20 December 1943. Original with Chaplain (Colonel) Robert Nay.

[19] U.S. War Department, *Commissioned Officers: Officers Appointed in the Army of the United States*, Army Regulation 605-10 (War Department, December 10, 1941) 13.c.

[20] U.S. Forces, European Theater, *Report on the Army Chaplain in the European Theater*. (Study 68, Bad Nauheim, 1945), Chapter 2.

[21] *History of USACHS: France 1918-Havard University*, 17.

[22] Roy J Honeywell, *Chaplains of the United States Army*, (Washington, DC: Office, Chief of Chaplains, 1958). 247.

[23] U.S. War Department, Technical Manual 16-205, *The Chaplain*. (War Department, July 5, 1944), 18.

[24] Parker C. Thompson, *From It European Antecedents to 1791: The United States Army Chaplaincy*

Volume I, (Office of the Chief of Chaplains, Department of the Army, Washington D.C., 1978), 113.

[25] Honeywell, 244.

[26] Gushwa, 109-110.

[27] Robert R. Venzke, *Confidence in Battle, Inspiration in Peace: The United States Army Chaplaincy 1945-1975 Volume V*, (Office of the Chief of Chaplains, Department of the Army, 1977), 52-53.

[28] Some books wrongly call this location at the time Lackland Field. San Antonio Aviation Cadet Center was an independent organization located on Kelly Field. On February, 3, 1948 the independent organization and location became Lackland Air Force Base.

[29] Honeywell, 250.

[30] Circular Letter 208, Office of the Chief of Chaplains, October 1, 1940.

[31] William D. Cleary, "How Army Chaplains Are Trained," in *Faith Of Our Fighters*, 106-107.

[32] William R. Arnold, "Message from our Chief of Chaplains," *The Army Chaplain: Official Publication of the Chaplains' Association of the Army of the United States, Vol VIII, No 3* (January 1938), 82.

[33] Honeywell, 251.

[34] Headquarters Army Service Forces, United States Army Chaplain School, *Manual For Instructors*, (Headquarters U.S. Army Chaplain School, Harvard University, Cambridge, Massachusetts, February, 1943), paragraph 58.

[35] Army Forces Pacif c, *History of Chaplains' Activities in the Pacific*, (Chaplain Section, Ground Headquarters, Army Forces Pacif c, 1946), 316.

[36] *Manual For Instructors,* paragraph 58.

[37] Ibid., Figure 9.

[38] Milton O. Beebe. Applicant For Appointment As Chaplain, United States Army, Form 399. (Beebe personnel file, located at the Army Chaplain Museum, Fort Jackson, SC).

[39] Milton O. Beebe. The Chaplain's Personal Obligations. *The Army Chaplain* Volume X Number 2 October-November 1939 Julian Yates Ed. (Chaplains Association of the Army of the United States, December 1939) 50.

[40] Memorandum to Chaplain Leroy R. Priest from Chaplain (Major General) Luther Miller, War Department, Office of the Chief of Chaplains Washington, July 8, 1946.

[41] Gushwa, 110.

[42] Honeywell, 249.

Chapter 5

Materiel

The United States industrial might and capacity during World War II provided the means for OCCH to equip Chaplains to provide for the free exercise of religion. The Army defines materiel as:

> All items (including ships, tanks, self-propelled weapons, aircraft, and so forth, and related spares, repair parts, and support equipment but excluding real property, installations, and utilities) necessary to equip, operate, maintain, and support military activities without distinction as to its application for administrative or combat purposes.[1]

A suggested definition for religious support materiel is:

> All items (ecclesiastical and equipment and supplies including those in chapel facilities and in a combat environment, communication and transportation assets, administrative equipment and supplies but excluding real property such as chapels and family life centers) necessary to provide and advise religious support without distinction as to its application for administrative or combat purposes and fully integrates into all aspects of a unit's operation and function.

OCCH filled the materiel gap in four ways. First, OCCH took advantage of the United States industrial might and modified and increased the Table of Equipment for a "Chaplain's Outfit" to operate in the new mechanized warfare environment. Second, the Chief of Chaplains engaged with civilian religious groups and industry to fill the materiel gap. Third, the chaplain's flag and the government purchased kit shows the importance of supporting all faith groups. Finally, oversight of the procurement process from the vendor to the user is essential for quality religious support.

The Chaplain's Outfit: Tables of Equipment From 1923-1945

The Chaplain's Outfit was the term used by the Army to reference the items chaplains required to provide religious support for their soldiers. The Table of Equipment in 1923 authorized each chaplain a "Chaplain's Outfit" that consisted of one field desk—regimental [size, ed.], containing a portable typewriter; one folding organ; 300 songbooks, religious and patriotic; and a chest as a container for books. This collection of equipment remained almost the same until Chaplain Arnold assumed duties as the Chief of Chaplains.

In the Army, Chaplains require the "basic tables of allowances" such as tentage and communication equipment. However, Tables of Equipment include materiel items for religious support personnel that are essential for the free exercise of religion. Chaplains who say, "All I need is a Bible" neglect the requirements of more liturgical Christian faith groups or other faith traditions. Chaplains must have on hand or have the ability to quickly acquire religious support items for soldiers within their organization. When they do this, they are supporting the free exercise of religion.

Historically, civilian organizations contributed religious items to chaplains as they left their congregations to serve in the military as farewell gifts. Denominational headquarters also provided kits for their chaplains. For Christian Chaplains, items consisted of chalices that were sometimes marked by the congregation or organization, such as the Knights of Columbus. Communion kits consisted of a communion tray, ciborium, and paten. For Jewish chaplains, customized cases included candlesticks, Sefer Torah, yad, several kippahs, and tallits. Religious groups also provided respective religious literature.

World War I and the expeditionary requirements such as standardization, weight, sustainability, and durability required the government to provide for the religious accommodation for its soldiers. Due in part to the National Defense Act of 1920, which established for the first time a Chief of Chaplains, the Army slowly began to recognize the importance of equipping chaplains just as they equipped other professional services within the Army.[2]

In Circular Letter 208, October 1, 1940, OCCH was interested in issues of supply. "Chaplains who are unable to secure the Chaplain's Outfit (Cir.#4 OQMG [Office of the Quartermaster General, ed.], 1938) within a reasonable time after making requisition are requested to notify this office."[3] Circular #4, OQMG was not clear that Chaplains needed to request a typewriter in addition to requesting the field desk. Therefore, in the same Circular Letter 208, OCCH reminded Chaplains to "obtain portable typewriters to complete field desks by requisition through channels to the Corps Area Quartermaster. The requisition should show make of typewriter desired."[4] OCCH recognized the difficulty chaplains were having with the supply system and provided guidance to overcome the bureaucracy. Clarity and preciseness are essential in every instruction from leadership. Failure to provide it results often in insufficient outcomes and additional corrective publications.

In Circular Letter 223, April 15, 1941, the War Department approved the placing of the chaplain's outfit on each troop transport ship.[5] The same letter provided for the increase of Table of Allowances for hymnals for posts, camps, and stations. In a memorandum, July 1, 1943, from Chaplain Arnold to all chaplains, the Chaplain's Outfit added the Chaplain's Flag (Jewish or Christian), the 150 *Song & Service Books for Ship and Field*, as well as the newly developed Chaplain's Scarf (Jewish or Christian). The scarves, like the typewriter, were a separate request.

Figure 10. Jewish scarf (left) Christian scarf (right). Notice the extra Tzitzit (tassels) on the Jewish scarf. Image from the author.

The Chaplain's Scarf created several issues for Chaplain Arnold that centered on the separation of church and state. The reason why the Chaplain's Scarf was a separate request was because the Roman Catholic Military Ordinariate did not allow their chaplains to wear the scarf.[6] Catholic Priests could only wear authorized vestments in the performance of their duties. During the war, Catholic chaplains had to get approval from the Pope to wear camouflaged vestments in the field.[7] The scarf on the other hand represented too closely the issues of the relationship between church and state. The scarf, both Jewish and Christian had the national symbol of the coat of arms over the tablets of the law or the cross. Some thought this represented religious groups were subservient to the state. In his monthly circular letter for September 1944, Chaplain Arnold argued how the scarf is actually a symbol of religious freedom.

> THE COAT OF ARMS AND CHAPLAIN'S INSIGNIA ON CHAPLAIN'S SCARF: The question has come up as to the relative positions of the Chaplain's Insignia and the Coat of Arms on the Chaplain's Scarf. It is believed that they are in harmony with good heraldic concepts, the American democratic philosophy, and a practical evaluation of the issues concerned. In no sense is the question of the primacy of allegiance to the state suggested since the genius of our government rests on the principle of the separation of state and church and affirms the freedom of religion and conscience as a principle to be preserved and protected. These principles, embodying freedoms which are unalienable rights, are accepted by the democratic state only. Therefore, the chaplain's insignia, symbol of religion, surmounted by the coat of arms, symbol of the protecting state, seem entirely appropriate.[8]

Chaplain Arnold may have remembered similar issues when the Army Service Uniform became the standard. On the lapel of the service uniform, the U.S. is over the cross or the tablets. During the 1920s Joseph Clemens, one of the contributors to the first Army chaplain field manual, disagreed with this concept and would place the cross over the U.S. on his service uniform.[9] As Chaplain Arnold brilliantly pointed out, the Chaplain Scarf is an excellent example of religious freedom and toleration.

For chapels, the Table of Allowances dictated one *Army and Navy Hymnal* for every two seats. Also, for each chapel, the Service Command Chaplain provided the Altar Set (cross, crucifix, candlesticks, and vases), Altar Cover, Pulpit and Lectern Scarves, Communion Set, National Colors and Staff, and requisition procedures for Jewish, Catholic and Protestant Scriptures.[10] Including chapels in the Table of Allowances shows Chaplain Arnold's emphasis upon the equal importance of garrison ministry along with chaplain ministry with front line troops.

However, an issue must have quickly arisen where there were not enough chaplain's outfits for those in the field. Therefore, a Memorandum from the Chief of Chaplains to all chaplains provided additional guidance:

> When in the U.S. and on a U.S. post, there is no need for the outfit. When on maneuvers, you will sign for an outfit and then return it at the conclusion of the training. When overseas the T/E [tables of equipment, ed.] of the particular organization will apply and will include the following items only: regimental desk, portable typewriter, chaplain's flag and scarf.[11]

Arnold's guidance set a priority for the distribution of equipment while ensuring support for all ministry requirements throughout the Army. Finally, just over two years after Pearl Harbor and two months before D-Day, over 7,000 chaplains received the required equipment as mentioned in Circular Letter 283, April 1, 1944 "all chaplains should have their Chaplain's Outfit."[12]

In December 1943, in the Ordnance authorizations separate from the Table of Equipment, chaplains were to receive the Truck—1/4–ton, and Trailer—1/4–ton.[13] Now each organization that was authorized a chaplain would also have a jeep and trailer. Now every chaplain had the same basic equipment as all soldiers in the Army, with the exception of a weapon, so they could provide effective religious support. Chaplains having a jeep and trailer constituted a change from World War I. During World War I Chaplains had to borrow vehicles or horses to visit soldiers in the hospital or remote locations. Early in 1919, the Jewish Welfare Board showed the effectiveness of having transportation. They brought every Jewish Chaplain Rabbi overseas a Ford automobile, "doubling their scope of work…and making them the envy of all the chaplains in France."[14]

Chaplain Arnold and OCCH's familiarity with the Tables of Allocation allowed them to change and modify the requirement when needed. The ability to adapt required close monitoring, assessing, and providing clear instruction.

Civilian Religious Groups And Industry Filling The Gap

Many branches within the Army have their unique equipment and tools to accomplish their mission. Engineers, signal and communication soldiers, cooks, and medical personnel all must have equipment that can meet the requirement under harsh conditions over long periods. The chaplain's kit is the one unique piece of equipment for chaplains. The one unique piece of equipment consists of a collection of religious items used to perform religious services that support the free exercise of religion.

Before World War II, religious organizations would provide their deploying clergy with the necessary items for them to perform their religious functions. However, this led to a wide variety of items on the battlefield and excessive weight for shipment. Chaplain Arnold wanted to resolve the problems of excessive weight for shipment and the burden it put upon the religious support teams. In November 1941, OCCH suggested aluminum communion trays for troop units.[15] However, due to the need for aluminum for aircraft, industry used cheaper metals that were silver coated for the communion trays. During the war, the Army chaplaincy purchased fragile wooden communion trays until more durable metal communion trays entered the supply system.[16]

The history of the civilian wartime communion kit shows similar struggles with obtaining large quantities of wartime metals. OCCH overcame this obstacle by using the power and ingenuity of industry. Also, OCCH communicated and coordinated with religious organizations of different faith backgrounds while also obtaining feedback and information from chaplains in the field. Their effort to engage at every level leveraged the bonds between OCCH, the chaplain in the field, and the chaplain's denomination. The result was the greatest number of communion sets ever produced and contributed to the religious support requirements in World War II and the Korean conflict.

Early in January 1941, Chaplain William H. Nicolas of the 102nd Infantry Division and a veteran of World War I, brought several silver communion pieces he had used in World War I to International Silver Ecclesiastical Department of Meriden, Connecticut for refurbishment. International Silver fashioned a complete communion kit from the Nicolas's communion pieces, including a special case. Other chaplains requested similar items, and International Silver experimented with several prototypes. In June 1941, the Episcopal Church in the United States of America was the first denomination to request kits for their chaplains. By early 1942, Episcopal, Lutheran, Presbyterian, Congregational, and Catholic organizations were requesting orders from International Silver. Then to the surprise of the International's specialists in religious customs, "it was the first time the Baptists had used a communion set as part of their equipment."[17]Now, most of the denominations had a chaplain's kit.

These commercial communion kits complied with wartime requirements for weight and standardized markings on the outside. Many of the kits contained a large individual cup communion tray with 30 small glass individual cups. The rigors of transportation and wartime service broke many of these glass cups. OCCH ordered and provided glass replacements[18] or small paper cups[19] for the denominationally provided communion kits.

Figure 11. Sterling Silver Denominational Procured Chaplain's Kit
Source: The author's personal collection.

As mentioned earlier, obtaining the special metal for the communion trays presented a problem for the government and the manufacturer. The same applied to International

Silver. Ecclesiastical items were copper alloy overlaid with silver. The War Production Board recommended carbon steel instead of copper. International Silver, along with The War-Time Commission of the Presbyterian Church in the United States of America, The National Lutheran Council, Congregational Churches of the United States of America and the American Baptist Home Mission Society and several other religious groups, protested and provided an impact statement with justification. When the War Production Board met, it allowed the use of the copper alloy without restriction.[20] The Army Chaplaincy was not nor should it ever be the only organization to interface with religious groups. Religious organizations may express their requirements to civilian authorities when the requirement is not religious in substance.

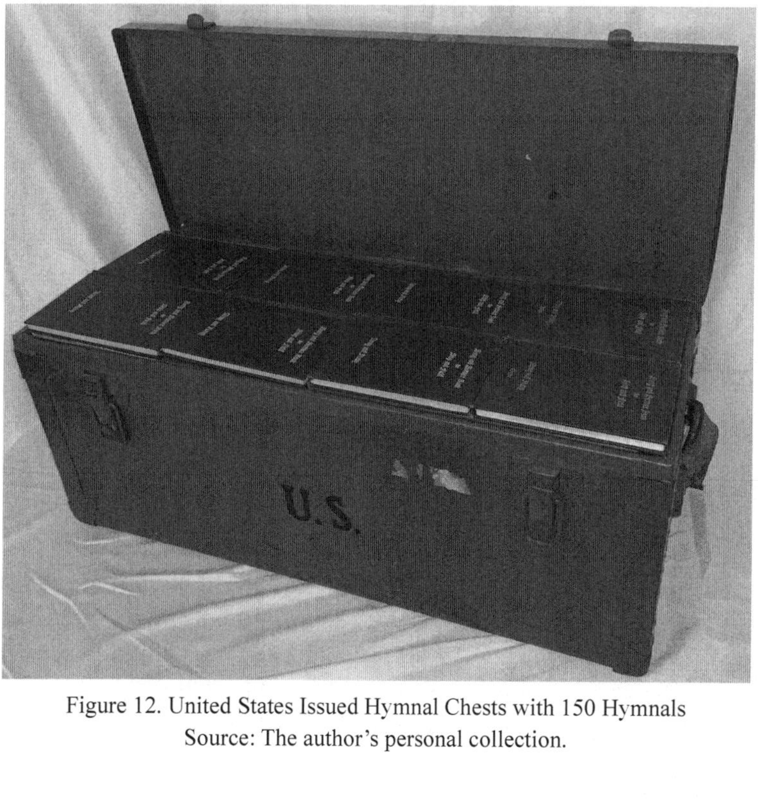

Figure 12. United States Issued Hymnal Chests with 150 Hymnals
Source: The author's personal collection.

In addition to attempts at minimizing equipment weight issues for communion kits, additional attention went into the development of the hymnal chest. An earlier version of the Hymnal Chest, Model 1921, could only hold forty-eight Army and Navy Hymnals. Although both the design of the hymnal and the chest allowed for the maximum number of hymnals with little wasted space, however, the size and the weight became impractical for large formations. Chaplain Ivan Bennett edited a smaller hymnal, *Songs and Service Book for Ship and Field: Army And Navy*, and oversaw the complete redesign of the hymnal chest that allowed for 150 *Songs and Service Book for Ship and Field.*[21] The efficiency of this design resulted in the utilization of the hymnal chest and hymnals for World War II, and the Korean and Vietnam conflicts.

In some cases, particularly at the start of World War II, chaplains would use whatever they could find to perform their religious mission. One resourceful chaplain cut down 50-caliber shells and used them for communion cups.[22] While religious and communion services were memorable and reflected Micah 4:3 "They shall beat their swords into plowshares and their spears into pruning hooks" the message that the military was not providing for the religious requirements for service members could prove to be disastrous for civilian morale and overall recruiting efforts.

Chaplain Arnold addressed this problem by obtaining a budget from the War Department. In addition to the advancement regarding the Table of Equipment, Circular Letter 215, January 15, 1941, mentions a budget of $53,600 from the War Department for the "Chief of Chaplains' Religious Fund."[23] This fund amounted to approximately $40.00 for each chaplain on active duty to procure religious supplies and materials.[24] Each chaplain could use this funding for religious reading materials, religious equipment, to include the portable reed pump organ, and the "portable altar" chaplain's field kit. The fund did not support the construction or procurement of bulletin boards or other basic items.[25] The unit normally provided these items. In 1943, the Chief of Chaplains Religious Fund was renamed Religious Activities Fund. In 1944, the Religious Activities Fund replaced the abolished unit chaplain's fund. The intent was to allow better accountability of the sacred offerings and local funding to support post wide initiatives.[26]

Before the establishment of the Chief of Chaplains Religious Fund, denominations provided chaplains with small portable communion field kits. In Technical Circular 7, May 1, 1942, "Chaplains would be wise to inquire at their denomination headquarters." The circular provides a six-page list of the contact information for the denominations.[27] However, if the chaplain did not have a communion kit, funds were available for the chaplain to purchase one. The Chaplain would have to provide a letter stating that his denomination does not "furnish him with a similar kit."[28] The government procured communion kit would remain with the unit and not the chaplain, while the denominational supplied kit remained with the chaplain for the duration of the war.[29]

Chaplain Arnold did not want to entirely "turn off" the supply chain from civilian organizations. On the one hand, he desperately needed the items for his chaplains. "It is not intended that churches making this gift discontinue the practice."[30] The budget from the War Department and contract process to procure a chaplain's kit could not meet the demand. On the other hand, just as there were tire and metal donations, and victory gardens to gather the civilian population behind the war effort, so too denominations and religious groups contributed to the war effort.

Many faith groups provided literature and care packages for their soldiers through mailings or off post establishments. The Jewish Welfare Board provided for Jewish service members. The Confraternity of the Precious Blood, a Roman Catholic organization that produced, among many things, the Government issued Prayer book mentioned under the Doctrine domain, provided items for Roman Catholic Service members. The Greek Orthodox Archdiocese of North and South America developed a liturgical prayer book for the faithful.[31] Numerous Protestant publications such as *The Link* and denominational publications such as *The Lutheran Chaplain* provided devotional material for Protestant service members. The Church of Jesus Christ of Latter-Day Saints published *The Book of Mormon* and the *Principles of the Gospel* that was almost the same size and color as the brown Testament Scriptures provided for the Jewish, Roman Catholic, and Protestant service members, but without President Roosevelt's and Chaplain Arnold's letter inside. Religious organizations were not limited to just providing literature. Even the Military Ordinariate for Roman Catholic Chaplains offered an electric host maker in the event obtaining hosts was not possible.[32] Allowing religious groups to provide monitored religious support enhanced religious support overall and linked the soldier with his faith, family, and the community of the faithful within the United States.

The Citizens' Committee for the Army and Navy (CCAN) provides an example of the cooperative efforts by the Jewish, Roman Catholic, and Protestant religious groups. The CCAN, led by Junius S. Morgan, gathered several prominent artists and commissioned triptychs. These triptychs, three-paneled paintings for chapels or troops in the field were in two sizes. The five by eight feet triptychs served in large areas and buildings, and the three feet by five feet triptychs for small areas and troop units.[33] Between

December 1941 and December 1945, CCAN provided 554 triptychs for the Army and Navy Chaplaincy.[34] There were Jewish, Roman Catholic, and Protestant triptychs. One had the Archangel Michael trouncing upon the devil, and for the Jewish faith, the lions of Judah flank the Ten Commandments.[35] Even the famous African American Artist Ellis Wilson, who was part of the Harlem Renaissance, contributed several triptychs. One of his triptych's "Saint Benedict the Moor," portrays the Saint who was born of African slaves and who became a Franciscan friar in the 16th century.[36] The Citizens' Committee for the Army and Navy also distributed approximately 625,000 pictures throughout the world. These images were therapeutic in hospitals and effective in "creating an atmosphere of worship even in strident surroundings."[37]

Figure 13. Triptych Provided by the Citizens' Committee for the Army and Navy
Source: The author's personal collection.

The Chief of Chaplains engaged and utilized industry and civilian organizations to help fill the wartime crises materiel gap. The impact was the meaningful bond between soldiers and those who sent them to war.

Chaplain's Flag, Religious Sensitivities, and the Government Purchased Kit

The chaplain's flag is the first official piece of equipment produced for the Army Chaplain. The development of the flag and the first government purchased communion kit shows the importance to include all faith groups in the development of religious equipment for chaplains.

Before the National Defense Act of 1920, Regular Army chaplains had to write and submit an essay to the War Department for promotion to the next rank. Someone within the Army's Adjutant General section would read the paper and recommend or not recommend the chaplain for promotion. George J. Waring, a Roman Catholic Chaplain from the fields of Iowa, submitted his promotion essay in 1912. The Adjutant General liked Waring's manuscript so much it was passed from the War Department to the Government Printing Office as *Chaplain Duties And How Best To Accomplish His Work.*[38] In this pamphlet, Waring recommended a chaplain's flag with blue background and a white cross.[39] The War Department adopted his recommendation, and several Christian chaplains used the flag with a cross in World War I. Although there were Army Jewish Chaplains in Europe, the regulation and guidance was not clear during or after the war for Jewish Chaplains.

The 1923 Army Regulation 260-10, *Flags, Colors, Standards, Guidons*, describes the flag as "blue bunting, 2 feet hoist by 3 feet fly, with a white Latin cross, 1 foot

6 inches in height, with arms 3 inches in width, in the center. This flag will be used for field service only."[40] There is no mention of a Jewish Chaplain's flag even though twenty-two Jewish Chaplains served in the Army during World War I.[41] It was not until three editions later in 1942 that AR 260-10 added the Jewish Chaplain's flag. "The flag of a chaplain of the Jewish faith will be of blue wool bunting, 2 feet hoist by 3 feet fly, with a white double tablet bearing Roman numerals from I to X surmounted by two equilateral triangles interlaced [Star of David, ed,], outlined in blue, 14 inches in height, in the center."[42] The writers had a chance earlier in 1927 to add the Jewish Chaplains flag to the regulation. Instead, they merely stated the chaplains' flag will "designate the time and place of divine service, and in the field to indicate the chaplain's quarters or office."[43]

The creator of the Jewish chaplain's flag also should have been aware of Jewish religious and cultural history. To include Roman numerals instead of the English or Jewish numerals to convey the Ten Commandments on the tablets show insensitivity toward the Jewish faith and tradition for two reasons. First, Romans harshly ruled the Jewish people that culminated in the siege and destruction of Masada. Second, Jewish religious items from the early 20th century included Jewish numerical characters in their display of the Ten Commandments.

Competent and mindful acquisition personnel should always include all religious traditions when developing or overhauling religious equipment and facilities. Devoting a significant amount of resources toward a piece of equipment that excludes several faith traditions is not good stewardship, nor does it support the free exercise of religion. Further, the overhauling of religious equipment and in particular, religious facilities to meet one specific faith group could violate the establishment clause of the Constitution. For example, when one piece of equipment or facility's sole use for one distinctive religious group, it suggest endorsement of that group. An exception is when a cost-benefit analysis justifies spending funds for low-cost equipment such as sacred chalices for Roman Catholics, or prayer rugs for Muslim soldiers. In this situation, the faith tradition does not allow for ordinary cups or rugs in the practice of the faithful.

The World War II government purchased chaplain kit went through a similar situation regarding religious sensitivities. Early in 1943, OCCH purchased "a limited number of Portable Altar Sets…and distributed one to each Division Chaplain and one to each Corps Chaplain for use with his truck, and one to each Army [ed. echelon] Chaplain."[44]

Figure 14. United States Issued Kit
Source: The author's personal collection.

Figure 15. United States Issued Kit Compared with the Denominational Procured Chaplain's Kit. Source: The author's personal collection.

National Church Goods Supply Company in Philadelphia, Pennsylvania, received the contract and the requirements. Since each item, to include the case, became government property, each piece had the regulatory "U.S." stamp on it just like most equipment within the military inventory. The kit resembled a "portable altar" because the case would open, and an inner portion would fold out and convert into an altar.[45] The maroon cloth would cover the lid and the folded portion of the case. The contents included two candlesticks, two glass cruets, chalice, intinction cup, paten, host plate, eight cup glass communion tray with lid, ciborium, and a unique combined cross and crucifix. To make this case usable for Roman Catholics and Protestants, on one side of the cross was the corpus of Jesus Christ and on the other side was IHS. The intent was that if a Roman Catholic or some liturgical chaplains received the kit, they would turn the crucifix with the corpus toward the congregation. When a Protestant chaplain used the cross, the IHS would face the congregation.

Figure 16. United States Issued Offering Plate with U.S. stamp.
Source: The author's personal collection.

Many Roman Catholic and Protestant Chaplains did not like this kit. For the Roman Catholics, the chalice is a consecrated item and it must remain strictly in the service of the Roman Catholic Mass. Therefore, Protestant chaplains could not use the chalice. Word of this must have gotten back to the Chief of Chaplains who wrote in January 1944, "Neither the entire set nor any of its component parts may be consecrated by the chaplain for his sole use since this action would unnecessarily restrict the use or the set by chaplains of other faiths."[46] The new communion kit provided a dilemma for Chaplain Arnold, who knew Roman Catholic requirements. Perhaps Chaplain Arnold hoped that his Roman Catholic chaplains would use the sacred chalice provided by the sending congregation or the Knights of Columbus along with the Army issued kit.

The Army issued kit was objectionable to several Protestant Chaplains because the corpus remained on the reverse side of the cross. The multi-purpose functionality of the crucifix/cross conflicted with most Protestant soteriology, the doctrine of salvation. Therefore, many chaplains unscrewed the three screws that kept the corpus attached to the cross and removed it for their services, and many were never reattached. When the Chief of Chaplains issued the next version of the kit in 1951 during the Korean War, the corpus had a clip on the corpus that could easily attach and detach from the cross. Also by this time, the Army provided a single chalice with paten in a small case.

Oversight of the Procurement Process: The Reed Pump Organ.

To guarantee the materials contributed to effective religious support, OCCH monitored the procurement of all supplies. As mentioned earlier regarding the contract with the Roman Catholic readings provided by the Confraternity of the Precious Blood and the development of the chaplain's flag, it was imperative that careful attention to the religious content in all the items conformed to the free exercise of religion and all things that enhance religious worship.

OCCH had an initial budget of $53,600 from the War Department for the "Chief of Chaplains' Religious Fund." [47] In a memorandum dated July 7, 1941, "All requests for purchases from this fund are to be made through OCCH in order to be assured of uniform supplies, better prices and a suitable method of accounting."[48] The impact of the careful oversight in the acquisition of the chaplain field organ for the Chaplain's Outfit shows the importance of this oversight.

The reed pump organ in Circular #4, OQMG, was the Bilhorn reed organ. Bilhorn had the government contract;

however, the product was inferior to the Estey pump organ from the company in Brattleboro, Vermont. Estey initially was unwilling to compromise the quality of its design to meet the weight requirements dictated by the War Department. Eventually, the lean years of the Depression era and the opportunity to have a significant contract prompted Estey to conform slightly and lower the quality of their organs to meet the War Department's requirements.

Pursuant to Directive P-E-67, dated December 11, 1940, the Army purchased from Estey 1,200 organs for $41,191. Although the Estey organ was much better than the Bilhorn, transportation over water and harsh climates resulted in significant problems, such as several keys playing at once, or not playing at all.[49] Chief of Chaplains Circulars during this time acknowledged the problems and provided advice on repair issues. Circular Letter 256, August 1, 1942, included an enclosure from Chaplain Henry Fairman, 136th Infantry, and a "reed repair man," who provided several pages of repair advice.[50]As soon as there was any concern raised about the functionality of the Estey organ, OCCH issued information and guidance. Even when the correction and input came from a chaplain at the lowest echelon, the Chief acknowledged him and shared his expertise.

The oversight of the performance of equipment also meant OCCH had to screen feedback from the battlefield. In May 1943, the After Action Report from the Chaplains of the 7th Infantry Division who served in the Battle of Attu recommended the use of the accordion. Chaplain (Lieutenant Colonel) Reuben E. Curtis the Division Chaplain wrote, "The field organ proved to be very inadequate and unsatisfactory, as it is not strong enough to stand the rigors of the field, neither will it operate in the continued damp weather we have. The accordion was successfully substituted as a musical instrument."[51]

Figure 17. Estey Military Pump Organ.
Source. The author's personal collection.

The Chief of Chaplains already considered and rejected the accordion as the official government purchased musical instrument for chaplains in the field. First, the environmental conditions of Alaska were extreme. Most equipment would not have to stand up to these conditions. Second, during the Tennessee Maneuvers in the summer of 1941, just before the war, chaplains or their assistants would use their "piano accordion" for worship or sing-alongs.[52] A chaplain who served on a transport ship in the Pacific had “a Jewish lad with his piano accordion” play hymns and “some good music” after the worship service.[53] The case for these accordions protected the instrument from the elements. However, the number of soldiers who could play the accordion compared with the pump organ made the organ the preferred and authorized instrument.

The feedback from the chaplains and the lessons learned regarding the design of the organ, assisted Estey in producing in September 1945 the M-45 Folding Chaplain’s Organ with Case. During transportation, this organ rested in a rubber-sealed case. Metal legs replaced the heavy wooden frame. Only a limited number of M-45 Folding Chaplains Organ saw production due to the war ending in September 1945. In order to address lingering maintenance issues, OCCH formally produced Technical Manual 10-750, *Chaplains' Folding Organ*, which included tips to repair the Estey and the M-45 pump organ.[54]

Another example of monitoring and assessing feedback centered on the Regimental Field Desk for the Chaplain’s Outfit. In most cases, the lowest echelon the chaplain was assigned was the regiment. Therefore, the assumption was that the chaplain needed the large regimental size field desk. This field desk required two people to carry it. However, feedback from the field during the war and after the war showed that chaplains asked for the smaller “Company Field Desk” that was less than half the size of the Regimental Field Desk.[55] After the war, the Army issued the Regimental Field Desk, the M-45 with metal handles. This desk was the same size and weight of the World War II desk but with metal handles because the leather handles on the earlier version would break due to the weight. The Chaplaincy did not consider the weight issue and the feedback from the After Action Reports. The post-World War II Chaplain Kit Table of Equipment continued to require the larger Regimental Field Desk.

OCCH guided the materiel requirements for Chaplains to provide religious support to all denominations without discrimination. The guidance enabled chaplains to fulfill their sacred calling in every echelon wherever they served. Also, the standardizing of materials within the Army system and providing quality assurance placed the chaplaincy on equal standing with other professional branches within the Service.

Notes

[1] U.S. Department of the Army, Army Regulation 5-22, *The Army Force Modernization Proponent System*, (Washington, DC: U.S. Department of the Army, October 28, 2015), 13.

[2] Roy J Honeywell, *Chaplains of the United States Army,* (Washington, DC: Office, Chief of Chaplains, 1958), 254.

[3] Headquarters Army Service Forces. Circular Letter 4, Office of the Quartermaster General, October 15, 1941.

[4] Ibid.

[5] Circular Letter 223, Office of the Chief of Chaplains, April 15, 1941.

[6] U.S. Forces, European Theater, *Report on the Army Chaplain in the European Theater*. (Study 68, Bad Nauheim, 1945), 106.

[7] Military Ordinariate, Circular Letter 31, (462 Madison Ave NY, NY, April 22, 1943.

[8] Circular Letter 288, Office of the Chief of Chaplains, September, 1944.

[9] The author has in his collection several photos of Joseph Clemens with the cross above the U.S. on his uniform.

[10] War Department Army Services Forces, Office of the Chief of Chaplains, July 1, 1943, Memorandum to All Chaplains.

[11] War Department Army Services Forces, Office of the Chief of Chaplains, August 31, 1943, Memorandum to All Chaplains.

[12] Circular Letter 283, Office of the Chief of Chaplains, April 1, 1944.

[13] Circular Letter 279, Office of the Chief of Chaplains, December 1, 1943.

[14] David Max Eichhorn, A History of the American Jewish Military Chaplaincy, *Rabbis in Uniform: A Century of Service to God and Country 1862-1962,* Louis Barish editor (Jonathan David: New York, 1962), 10.

[15] Circular Letter 238, Office of the Chief of Chaplains, November 1, 1941.

[16] Circular Letter, 282, March 1, 1944.

[17] Earl Chapin May, *Century of Silver, 1847-1947, Connecticut Yankees and a Noble Metal*, (Robert M. McBride & Company, New York, 1947), 268.

[18] Circular Letter 286, Office of the Chief of Chaplains, July 1, 1944.

[19] Circular Letter 282, Office of the Chief of Chaplains, March 1, 1944.

[20] May, 270.

[21] https://www.washingtonpost.com/archive/local/1980/06/18/maj-gen-ivan-bennett-chief-of-army-chaplains/f89923c7-f7ce-4f9f-98fe-cf9b402100c6/ (accessed September 2, 2018).

[22] Honeywell, 256.

[23] Memorandum War Department Office of the Chief of Chaplains July 7, 1941 Subject: Chief of Chaplains' Religious Fund, To All Chaplains on Active Duty.

[24] Circular Letter 215, Office of the Chief of Chaplains, January 15, 1941.

[25] Circular Letter 252, Office of the Chief of Chaplains, June 1, 1942.

[26] Roy J. Honeywell, "Chaplains of the Armed Force: A Compendious History," *The Military Chaplain* Volume XXX, Number 4 (April 1957), 15.

[27] U.S. War Department, Technical Circular Number 7, (Office Chief of Chaplains, Services of Supply, War Department, Washington D.C. May 1, 1942), 11-16.

[28] Circular Letter 280, Office of the Chief of Chaplains, January 1, 1944.

[29] After the war, many denominations allowed their chaplains to keep the communion kits.

[30] Circular Letter 280, Office of the Chief of Chaplains, January 1, 1944.

[31] Greek Orthodox Archdiocese of North and South America, *Greek Orthodox Prayer Book—For the Use of Soldiers and Sailors on the United States of America*, (Greek Theological School Press), 1942.

[32] Military Ordinariate, Circular Letter 26, (462 Madison Ave NY, NY, October 27, 1942), 2.

[33] *Report On The Army Chaplain In The European Theater* (The General Board, United States Forces, European Theater, Study Number 68), 43.

[34] Monroe Drew, Jr., *Triptychs*, The Army and Navy Chaplain. (Volume XVII, July-August, 1946, Number 1), 34. Holly Mitchem, *Triptychs for Victory: Altar Paintings for the American Armed Forces in World War II*, soon to be published Ph.D. thesis places this number at 472 triptychs. Her reference is the CCAN files.

[35] Circular Letter 246, Office of the Chief of Chaplains, March 2, 1942.

[36] Elizabeth McCracken,*Triptychs for Servicemen*, The Living Church (Morehouse-Gorham Co., January 30, 1944), 13-14.

[37] *Report On The Army Chaplain In The European Theater*, 43.

[38] George J Waring, *Chaplain's Duties and How Best to Accomplish His Work*. (Washington, DC: GPO, 1912).

[39] Ibid., 16.

[40] U.S. War Department, Army Regulation Number 260-10. *Flags, Colors, Standards, And Guidons*, (War Department Washington, February 8, 1923), 17.

[41] David Max Eichhorn, A History of the American Jewish Military Chaplaincy, *Rabbis in Uniform: A Century of Service to God and Country 1862-1962,* Louis Barish editor (Jonathan David: New York, 1962), 10.

[42] U.S. War Department, Army Regulation Number 260-10 Change 3. *Flags, Colors, Standards, And Guidons,* (War Department Washington, August 18, 1942), 9.

[43] U.S. War Department, Army Regulation Number 260-10 Change 1. *Flags, Colors, Standards, And Guidons,* (War Department Washington, July 2, 1927), 3.

[44] Circular Letter 267, Office of the Chief of Chaplains, February 1, 1943.

[45] Army Quartermaster supply manuals referenced this item as "portable altar."

[46] Circular Letter 280, Office of the Chief of Chaplains, January 1, 1944.

[47] Memorandum War Department Office of the Chief of Chaplains July 7, 1941 Subject: Chief of Chaplains' Religious Fund, To All Chaplains on Active Duty.

[48] Ibid.

[49] Robert F. Gellerman, *The American Reed Organ and the Harmonium*, (Vestal Press, January 1, 1997), 146-147.

[50] Circular Letter 256, Office of the Chief of Chaplains, August 1, 1942.

[51] Headquarters, U.S. Army Troops, Office of the Chaplain, CONFIDENTIAL, *Report Activities of Chaplains, 7th Infantry Division, Battle of Attu,* (Headquarters, U.S. Army Troops, Office of the Chaplain, July 5, 1942), 3.

[52] Circular Letter 232, Office of the Chief of Chaplains, August 15, 1941. Circular Letter 238, Office of the Chief of Chaplains, November 1, 1941.

[53] "Author Not Provided," *Worship In The Jungle*, The Army and Navy Chaplain. (Volume XIV, October-November, 1943, Number 2), 34.

[54] U.S. Department of the Army, *Chaplains' Folding Organ*, Technical Manual 10-750, (Washington, DC: U.S. Department of the Army, February 1948).

[55] *Report On The Army Chaplain In The European Theater*, 102.

Chapter 6

Leadership and Education

Leadership and education is the cheapest domain in terms of dollars required. However, the costs increase significantly for the career soldier. The Army defines leader development and education as "the product of a learning continuum that comprises training, experience, formal education, and continual self-improvement."[1] A suggested definition for religious support is "leader development and education that strengthens and builds upon the foundational vocational calling of the chaplaincy and chaplain identity." During World War II, Chaplain Arnold shared his vision through bimonthly and monthly Chief of Chaplains Circular Letters, and publications, notably the Technical Circular Letters. Secondly, Chaplain Arnold's leadership stressed the importance of the "calling" of civilian clergy to the Army Chaplaincy in almost everything he did. This included his oversight in the creation of the "*Soldiers of God: Official Chaplains March*" song.

Shared Vision and Shared Understanding: Military Documents and Publications

Chaplain Arnold's leadership in emphasizing the call of professional clergy who now serve as chaplains of the nation required similar emphasis upon education. In addition to the chaplain school, Chaplain Arnold provided a stream of information that facilitated personal self-development.

In the National Defense Act of 1920, Congress assigned proponency of each branch of the Army to a chief of the branch within the War Department. For the first time, the Chaplaincy had its own permanent head, the Chief of Chaplains. Institutionally, this new named official and his office naturally followed the example of other branch chiefs, including the requirement to publish circular letters. Chaplain Arnold used these circular letters to strengthen the bonds within the Chaplaincy by sharing information and providing guidance. Secondly, he used these letters to inform the public regarding the Army's efforts to care for the men and women in the Army.[2]

Chaplain Arnold wrote over three times more pages of circular letters than all his predecessors combined. Every chaplain and his commander received a copy of the circular letter. Also, as part of the public relations campaign, every religious organization that had a representative within the press corps, received a circular letter. During the war, Chaplain Honeywell, the Operations Chaplain in OCCH, wrote that one of the circular letters with enclosures weighed over a ton.[3] The information contained in these letters are War Department and Chief of Chaplains policies and directives, patriotic messages about the importance of freedom and duty against totalitarian regimes, professional career information, advice for performing ministry in a military setting, good news stories from the press about how chaplains were caring for the religious needs of their soldiers, book reviews, and personal information about births and deaths within the Chaplaincy. Finally, these letters contained encouraging words from the President, general officers, regimental commanders, and even privates.[4]

A key strategy for Chaplain Arnold for the circular letters consisted of building the bonds between chaplains at every echelon. On January 11, 1938, in his third Circular Letter after becoming Chief of Chaplains, he invited all chaplains to Washington D.C., for a training conference. To generate enthusiasm for this event, Chaplain Arnold included for the first time in a circular letter, a drawing of a smiling chaplain with a suitcase with the words, "I Wouldn't Miss It! Come On!! Let's Go!!!"[5] By connecting chaplains with one another, OCCH received information from chaplains and used these letters to get information out to chaplains wherever they may be.

WAR DEPARTMENT
OFFICE OF THE CHIEF OF CHAPLAINS
WASHINGTON

Circular Letter
No. 186, C.of Ch.

April 11, 1938.

A PRE-CONVENTION REMINDER. ATTENDANCE GOAL - 400 - Perhaps More.
The Answer is YOU.

WHY TO THE NATION'S CAPITOL?

TO THE 13TH ANNUAL CONVENTION OF THE CHAPLAINS' ASSOCIATION OF THE ARMY OF THE UNITED STATES — AT THE RALEIGH HOTEL - MAY - 17-18-19 - 1938

I WOULDN'T MISS IT!

General outline of program:

Tuesday --- Subject: "Worship in the Service."
Wednesday - Subject: "Sunday Worship" presented under the following:
(a) Gathering a Congregation; (b) Accessories and equipment for worship; (c) Preaching that appeals; (d) Sunday night groups.
Thursday -- Subject: "Week Day Worship" presented as a follow-up, as follows:
(a) With individuals; (b) With the sick; (c) With problem cases; (d) With week night services.

Note: Speakers of note from Church and State will bring messages you will want to hear. This will be the last convention in Washington for many years to come. Program specially arranged for visiting points of interest.
Attention is invited to Par. 2 h (3) AR 210-70 regarding detached service.
Inactive duty credits have also been authorized for all Reserve chaplains in attendance at the sessions.

COME ON !! LET'S GO !!!

Office Chief of Chaplains.

Figure 18. Example of the Monthly Circular Letter.
Source: The author's personal collection.

An example of how these circular letters strengthened the bonds of chaplains occurred in the latter half of 1941. In December 1940, Chaplain Aryeh Lev was the first Jewish Chaplain assigned to OCCH.[6] In July 15, 1941, Circular Letter 230, Chaplain Lev attached a letter with the following, "The Chief of Chaplains would be pleased to receive actual copies or even more preferably outlines of one or two of your best sermons which you have preached while on active duty."[7] Chaplain Lev wanted these documents by August 15, one month after the publication of the letter. Chaplain Lev received over 500 outlines and selected just 21 outlines and sermons for inclusion in a publication titled *What Chaplains Preach*.[8] This publication went to every chaplain and the church press corps with Circular Letter 247, March 16, 1942.[9] Even in terms of today's automation age, the resounding response was impressive and highlights the desire chaplains had to share knowledge and information to other chaplains. Chaplains wanted this information so they could enhance the religious support they provided for soldiers. Further, soldiers tend to contribute more in sensing sessions, town halls, and surveys, when they see feedback from their input. Soldier input does not necessarily mean that soldiers get what they want, but if they know leaders consider their feedback, they will invest and trust in the leadership and the organization.

Figure 19. Chaplain (Captain) Aryeh Lev. Arlington Chapel, 1941.
Source: Arlington Chapel dedication memento.

As the war progressed, the information and instruction evolved within the circular letters. There were a few reprints of previous entries, such as how the chaplaincy fits into the Army organization (see Appendix 3). Letters always included encouragement about the effectiveness of chaplains upon the soldiers and appreciation from senior leaders about chaplain ministry. The themes of democracy and freedom were always present. Starting with Circular Letter 265, January 2, 1942, the technical and more sensitive information such as "Death Gratuity," Graves Registration, how to care for behavioral health issues, and how to write condolence letters, were an "Addenda" to the circular letter. Commanders and Chaplains only received the Addenda.

A key element of Arnold's success was keeping the command informed and providing information and guidance inside Army channels. The impact this had upon all soldiers is in the following letter to the Chief of Chaplains.

> May I say that the emphasis placed in your circular letters upon a warm personal religious experience is deeply appreciated. The officers and men have their opinion of what a chaplain should be, and when I find that view is repeatedly pressed upon me, I turn to the circulars as a means of refreshing my vision from a military viewpoint. There the consistency between Army life and a strong spiritual ministry is clearly declared. That is a big help.

The circular letters reminded the commander what chaplain duties required. Soldiers within his unit had a vision of what a chaplain does; however, Arnold's circular letters reinforced expectations. The Commander then took those words and encouraged his chaplain in front of his men providing that strategic message and guidance to the tactical level. Chaplain Arnold continually set the standard for religious support by reminding commanders how to lead their chaplains.

In reflection, the Army Chaplain European Theater of Operations After Action Report wrote the following.

> The Office of the Chief of Chaplains published a lithoprinted monthly circular which is distributed to all chaplains and the American religious press. A supplement, called the Addenda, is sent [*sic*] *with* the Circular Letter to chaplains alone and has the force of a technical directive from the Chief of Chaplains. The Circular Letters and Addenda served a real need and were generally well received.[10]

Over 75 percent of the chaplains in the European Theater of Operations found the circulars "worth it."[11]

In addition to the monthly circular letters, The Technical Circulars, *Helpful Hints and Other Information*, and initiatives by Senior Chaplains contributed to a learning organizational culture. Starting on August 1, 1941, and ending on November 6, 1942, OCCH published ten Technical Circulars. Each circular provided current directives, guidance, and suggestions for each topic. The topics in order were: Chapels, Equipment, Administration and Use; Division Chaplains; The Chaplain at the Reception Center; The Regimental or Unit Chaplain; The Chaplain in the Replacement Training Center; Marriage and Related Subjects; Chaplains Equipment; The Chaplain in the Hospital; The Service Command Chaplain; and Chaplain Assistants. These topics were most relevant to chaplains and provided a quick win for Chaplain Arnold to inform and equip the new chaplains.

In May 1942, OCCH published 67 pages of *Helpful Hints and Other Information*.[12] This document started with an index of nearly 500 terms with paragraph references for quick access in the document. Topics included dietary requirements, Sabbath observance requirements, how to wear the uniform and graves registration. Future wartime commandant of the Chaplain School, Chaplain (Colonel) Maurice W. Reynolds, the IV Army Corps Chaplain, followed Chaplain Arnold's example and published his *Practical Duties of an Army Chaplain*.[13] This document provided additional suggestions based upon directives and guidance in order to provide religious support to soldiers.

The circular letters and the technical letters were "quick wins" for Chaplain Arnold. For more enduring purposes, Chaplain Arnold directed a new Chaplain's manual, Technical Manual 16-205, *The Chaplain*, War Department,

July 5, 1944.[14] Previous manuals were just slight revisions of the original 1926 edition, Technical Manual 2270-5, *The Chaplain: His place and duties*.[15] A fresh, new manual addressing current doctrine and methodologies provided invaluable guidance to civilian clergy who transitioned to the Army chaplaincy.

Several other publications provided regulatory information and guidance for chaplains. Army Regulation 60-5, *Chaplains*, underwent four different versions between February 1941 to December 1944. Army Regulation 55-355, *The Transport Chaplain*, provided regulatory guidance for an Army chaplain serving a troop transport ship. In Technical Manual 10-630, *Graves Registration*, War Department, September 23, 1941, Chaplain duties were inserted. Early in the war, the government did produce the *Instruction Manual For Hammond Electric Organs*, War Department, 1942. In this manual, the Chief's office highlighted the need for routine maintenance by making "requests to commanding officers every six months for radio electricians to inspect the amplification mechanism as suggested on page 42 of the manual." In addition to monthly maintenance, chaplains had to "warn organists to be particularly careful that switches on the organs are turned off after use."[16] The Estey pump organ required numerous technical instructions in the circular letters. Finally, the Estey pump organ had its own government publication, Technical Manual 10-750, *Chaplains' Folding Organ*, Department of the Army, 1948.

Chaplain Arnold shared his vision for the Army chaplaincy through Army publications and documents. He wanted and needed information, advice, and more importantly, inspiration on how to provide religious support. Chaplain Arnold's leadership example and his dedication to providing educational materials equipped chaplains with what they needed to offer effective and efficient religious support. A residual effect of his leadership and efforts reinforced the chaplains' call to the chaplaincy by instilling pride in the ministry of the Army chaplaincy.

Soldiers Of God: Official Chaplains' March

Chaplain Arnold's leadership in the development of the *Soldiers of God: Official Chaplains' March*, reveals his ability to understand the importance of music in bonding Army chaplains toward a sense of comradery and a common goal. He did this by expressing his vision and empowering experts in music to develop the official chaplains march.

Music in worship has been and continues to be an essential element of worship. In 1920, the development of the *Army and Navy Chaplain Hymnal* was an example of music's importance in worship and within the chaplaincy.[17] Earlier in his career, Chaplain Julian E. Yates, the third Chief of Chaplains (1929-1933), oversaw and edited the first Army and Navy Hymnal. Yates, reflecting at the development of the hymnal saw it as "trials and tribulations" and "Possibly some historian[18] will tell the story in greater detail at a later day."[19] During war, music tells stories of old and inspires new generations. An example of one such song early in World War II was *Praise The Lord and Pass the Ammunition.*[20] This song tells the famous story of Navy Chaplain Howell M. Forgy aboard the USS New Orleans at Pearl Harbor on December 7, 1941.

During World War II, Chaplain Arnold delegated the editing of two hymnals to the hymnologist Chaplain Ivan Bennet. Bennet developed one hymnal for use in the chapel, and another hymnal for use on ship and in the field. Later in his career, Chaplain Bennett became the eighth Chief of Chaplains (1952-1954).[21] Music is very much a part of the chaplain in their calling to provide worship.

Chaplain Arnold felt the Army Chaplaincy needed their own official song to establish pride in their calling. In Circular Letter 281, February 1, 1944, Chaplain Arnold wrote:

> The Air Corps, Signal Corps, Artillery and Infantry all have had an official song representing the respective branch or service. Now the Chaplain Corps to has an official song titled SOLDIERS OF GOD. The spirit-stirring music, composed by Ben Machan, is a catchy melody in thrilling march time. The lyrics, by Private Hy Zaret, are appropriate for a song that will be known everywhere as the Official Chaplains' March. The Music Section, Special Services Division, upon request of this office, asked a number of song writers to submit compositions. From these the song, SOLDIERS OF GOD, was selected and approved by the Chief of Chaplains as the Official Chaplains' March. A preliminary arrangement of the song is included in the January 1944 issue of HIT KIT. The final approved version is being published by Remick Music Corporation.[22]

The "March" music would have resonated with chaplains at that time. For two and a half hours a day, chaplains practiced drill, close order (marching) at the Chaplain School. The 1943 War Department training film, *For God and Country,* included chaplains marching at the Chaplain School. As we mentioned before, in the movie an actor who would later become United States President, Ronald Reagan portrayed the role of Father Michael O'Keefe. In one scene at the Chaplain School, he lamented how his "body is suffering" from the previous scene of marching.[23]

However, the first edition of the *Soldiers of God* did not meet the intent of Chaplain Arnold.[24] In a memorandum from OCCH, Chaplain Arnold provided additional guidance.[25] First, in all future publications, the word "call" would be in quotes. Secondly, in addition to "Fighters of the nation, Everywhere at a thousand battle stations," an added verse of "Chaplains of the nation, Everywhere with our fighting congregation" was added and placed before Zaret's edition. Also before Zaret's edition an introductory chorus, "Faithful to God, we're serving on the battlefield today..." emphasized the divine providence of their calling even "on the battlefield today."[26] (Appendix 7 shows a comparison of the two versions)

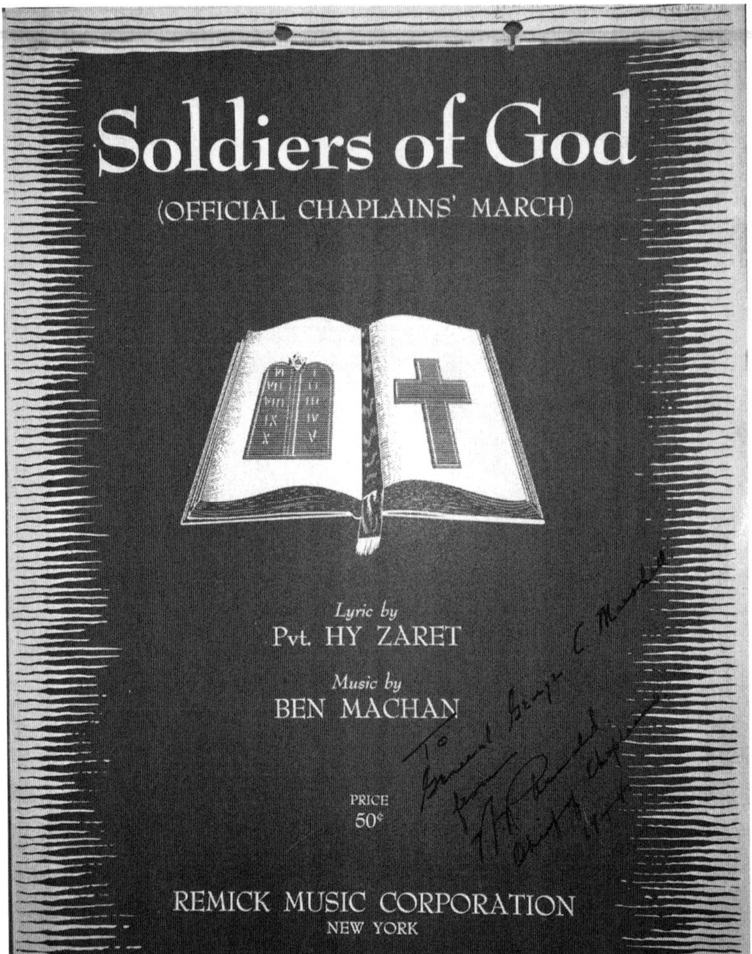

Figure 20. Soldiers of God Sheet Music given to General Marshal from Chaplain Arnold. Image from the author. Document from the George C. Marshall Library.

Chaplain Arnold's oversight and contribution to *Soldiers of God* enhanced the chaplain identity during World War II. Soldiers, Sailors, and Civilians reflected Arnold's vision for the march. In the memorandum and earlier publications of *Soldiers of God*, Arnold stressed that people have the option to choose to sing all of *Soldiers of God*, or just part of it. When Bing Crosby first sang the march at the Kraft Music Hall Radio Program on Easter, April 9, 1944, he only sang the first verse that included "Chaplains of the nation, everywhere with our fighting congregation." The adoption of this song influenced Navy chaplains who wanted their own song. However, the Navy never adopted the draft submitted in *The Army and Navy Chaplain* magazine.[27] At the close of the war, chaplains knew their calling as chaplains of the nation. On the second page of *History Chaplains' Activities in the Pacific,* compiled by the Chaplain Section Ground Headquarters, Army Forces Pacific (AFPAC), 1946, the same verses sung by Bing Crosby were included. Chaplains identified themselves with "Chaplains of the nation" and not entirely as "Fighters of the nation." The legacy of Soldiers of God reached Vietnam. General Creighton W. Abrams was a Field Grade Officer in World War II who became Commander, United States Military Assistance Command, Vietnam. In his introductory letter to a brochure that highlighted the ministry of chaplains in Vietnam, General Abrams referred to chaplains as Soldiers of God.[28]

Chaplain Arnold's circular letters equipped chaplains with reminders of their sacred calling. The increase in the quantity and quality of the circular letters and technical circulars emphasized to chaplains the knowledge and purpose of their calling. More importantly, circular letters reminded chaplains they are not alone. They remained bonded together with other chaplains based upon their sacred calling during the darkest and most difficult times of their sacred calling in the Army chaplaincy. In addition, chaplains knew they had the support and guidance of their senior chaplain leaders. Chaplain Arnold's oversight of *Soldiers of God* connected their civilian calling and their chaplain calling through music. The song told their story and reminded them they were Chaplains of the nation.

Notes

[1] U.S. Department of the Army, Army Regulation 5-22, *The Army Force Modernization Proponent System,* (Washington, DC: U.S. Department of the Army, October 28, 2015), 13.

[2] The numbering of the "Circular" letters within chaplain historiography is not always clear. Starting November 11, 1920 the Chief of Chaplains issued "Office Circular 1" to all chaplains and ended with Office Circular 135 November 24, 1925. In January 5, 1926 the OCCH started to name the documents "Circular Letter" and started again with the number 1 and published the last "Circular Letter 344" May 1, 1949.

[3] Roy J Honeywell, *Chaplains of the United States Army,* (Washington, DC: Office, Chief of Chaplains, 1958), 242.

[4] U.S. War Department, Memorandum To All Chaplains, Subject: Reports Pursuit to A.R. 827, Office of the Chief of Chaplains, November 11, 1920 – Department of the Army, Circular Letter 344, May 1, 1949. Original lithoprinted documents with Chaplain (Colonel) Robert Nay.

[5] Headquarters Army Service Forces. Circular Letter 186, Office of the Chief of Chaplains, January 11, 1938.

[6] Circular Letter 212, Office of the Chief of Chaplains, December 2, 1940.

[7] U.S. War Department Office of the Chief of Chaplains July 15, 1941 Subject Sermon Outlines. To: All Chaplains on Active Duty.

[8] Aryeh Lev ed., *What Chaplains Preach,* (Command and General Staff School, 1942).

[9] Circular Letter 247, Office of the Chief of Chaplains, March 16, 1942.

[10] *Report On The Army Chaplain In The European Theater* (The General Board, United States Forces, European Theater, Study Number 68), 21.

[11] Ibid.

[12] Department of the Army, *Helpful Hints and Other Information,* (Office of the Chief of Chaplains, May 1942) Original lithoprinted with Chaplain (Colonel) Robert Nay.

[13] Maurice W. Reynolds, *Practical Duties of An Army Chaplain*, (Chaplain, IV Corps, Unpublished, No Date) Original lithoprinted with Chaplain (Colonel) Robert Nay.

[14] U.S. War Department, Technical Manual 16-205, *The Chaplain*, (War Department, July 5, 1944).

[15] U.S. War Department, Technical Manual 2270-5, *The Chaplain: His place and duties*, (War Department,1926).

[16] Circular Letter 239, Office of the Chief of Chaplains, November 15, 1941.

[17] Julian E. Yates, ed., *The Army and Navy Hymnal*, (The Century Company, New York, 1920).

[18] One Chaplain did write a history of the hymnal. Gary W. Carr. *The Development of the Book of Worship for the United States Forces*, (A thesis submitted in partial fulfillment of requirements for the degree of Master of Theology in the Divinity School of Duke University, 1996).

[19] Julian E. Yates, The Hymnal. *The Army Chaplain* Volume X Number 3 January – February 1940 (Chaplains' Association Of The Army Of The United States, February 1940), 118.

[20] Frank Loesser, *Praise the Lord and Pass The Ammunition!!* (Famous Music Corp, 1942).

[21] In 1951, Chaplain Bennett served in the dual capacity as Far East Command Chaplain and the United

Nations Command Chaplain. Bennett edited two separate hymnals during the Korean conflict, *Psalms and Hymns and Spiritual Songs: Chinese-English*, and *Psalms and Hymns and Spiritual Songs: Korean-English*. Both hymnals included a Foreword by General Douglas MacArthur.

[22] Circular Letter 281, Office of the Chief of Chaplains, February 1, 1944.

[23] U.S. War Department, *For God and Country*. Training Film TF 16-2037, (U.S. War Department, Army Service Forces Signal Corps Production, 1943).

[24] Ben Machan and Hy Zaret, *Soldiers of God: Official Chaplains' March*, Hit Kit, (Remick Music Corporation, NY, 1943. Found in, Special Services Division, Army Service Forces, *Hit Kit*, January 1944).

[25] U.S. War Department, Army Service Forces, Office of the Chief of Chaplains December 16, 1943 To: Music Section, Special Service Division, Attention: Captain Harry Salter. Copy with the Harry Salter papers, Great American Songbook Foundation, 1 Center Green, Carmel IN 46032.

[26] Ben Machan and Hy Zaret, *Soldiers of God: Official Chaplains' March*, (Remick Music Corporation, NY, 1944).

[27] "Author Not Provided," *Songs of the Two Chaplains Corps*, The Army and Navy Chaplain. (Volume XVI, July-August, 1945, Number 1), 26.

[28] United States Military Assistance Vietnam, *Chaplains Vietnam*, no date.

Chapter 7

Personnel

Personnel is critical for any military. Recruiting and retaining the best Army chaplains is essential for an effective fighting force. The Army defines Personnel as "The development of manpower and personnel plans, programs, and policies necessary to man, support and sustain the Army."[1] A suggested religious support definition for Personnel is "Recruit, manage, and retain qualified religious support professionals in order to provide for the free exercise of religion in the Army."

The mobilization from a peacetime Army to World War II provides fruitful examples for future planners within the Army. Adjusting personnel requirements, interaction with religious groups and denominations, managing personnel deploying overseas, the procurement and quota system, facilitating African Americans as chaplains and religious lay leaders, and Chaplain Assistants contributed to a successful personnel function in order to provide personnel for religious support. As a foundation for this chapter, it is important to know the number of chaplains authorized and assigned before World War II.

Congress approved on May 5, 1918, the fixed number of chaplains at a ratio of one chaplain for every 1,200 officers and enlisted men, and twenty at large selected on a denominational basis, which would reflect approximately the size of the various religious groups in the United States.[2]

The 1920 National Defense Act authorized 125 Regular Army Chaplains and 1,000 Reserve Chaplains. The mobilization plans of 1925 called for a goal of 1,870 combined Regular Army and Reserve Chaplains. The chaplaincy never reached this goal during the interwar years. The number included the Depression era civilian-led Civilian Conservation Corps (CCC). Although the military and the chaplaincy endorsed and supported this effort as a means to practice mobilization and organization, the number never reached more than 200 chaplains within the CCC.

On September 22, 1941, Congress authorized appointments for the Army of the United States (AUS), which allowed temporary authorizations of chaplains based upon a national emergency.[3] These appointments extended beyond the regulatory number of authorizations for the Regular Army and Reserves. When the need for chaplains became exceedingly great for the Army, Chaplain Arnold lowered the standard for appointments to the AUS. The authorization of the AUS meant that following the emergency, those appointments could apply for the few Regular Army slots, or the Reserve, or National Guard.[4]

In the summer of 1940, there were 137 chaplains in the Regular Army. Records for the Reserve chaplains are sketchy, but of the 1,000 Reserve chaplains, 770 were eligible for active duty. Of the 770 Reserve chaplains, 145 of them were serving with the Army and about 100 with the Civilian Conservation Corps. By the day of the Pearl Harbor attack, 140 chaplains of the Regular Army, 298 of the National Guard and 1,040 Reserve chaplains were on duty, a total of 1,478.[5] The Army needed 8,000 chaplains based upon the projected number of soldiers needed for the war. From September 9, 1939, to September 2, 1945, a total of 9,117 chaplains served in the Army. When the war ended, the total was 8,141 chaplains on active duty (Jewish 243, Catholic 2,278, and Protestant 5,620).[6] Chaplain Arnold's knowledge, wisdom, and leadership made this possible.

Adjusting Personnel Requirements

In September 1940, Secretary of War General George Marshall asked Chaplain Arnold for a plan to meet the spiritual needs of the Army under the Selective Service Act. Arnold mentioned seven points that were to be of importance throughout the war: Procurement and Distribution of Chaplains; Training; Chapels; Cooperation with Church

Groups; Publicity; and Cooperation of Military Authorities. Chaplain Arnold also noted, "Procurement would be difficult but should be made on an equitable basis for the three major faiths."[7] Note that Arnold started with the personnel issue of procurement. Four of the seven points fall directly within the DOTMLPF-P domains, while the implied task of "cooperation" is within the other four domains. Moreover, in contrast with other individuals of the selective service, Chaplain Arnold did not request an exception to policy to draft civilian clergy for the Army chaplaincy.

Following the first general staff meeting after the attack on Pearl Harbor, Marshall wanted to see Arnold. In the meeting, Marshall told Arnold to report to him in a "day or two" to him with recommendations for chaplains. Arnold said, "I could do it now." He urged the appointment of chaplains in the ratio of one to eight hundred men.[8] Arnold did not get what he wanted, but Marshall certainly understood the wartime emergency and allowed one chaplain for every 1,000 soldiers under certain circumstances. Marshall's appreciation for chaplains and Arnold's competent and pastoral abilities allowed Marshall to trust Arnold and provide him with whatever he needed to accomplish the mission.[9]

Chaplain Arnold started to meet the wartime requirement for chaplains by adjusting the age restrictions for chaplains. Below are three points from a memorandum dated April 7, 1941. The points highlight the interaction of policy of age restrictions, and personnel requirement numbers.

> a. On April 3, 1941, the War Department announced the policy that certain age requirements in the Chaplain's Reserve would be relaxed in order to enable older chaplains, who are already commissioned, to apply for extended active duty for service other than with troop units.

b. The new ruling with respect to chaplains will permit the assignment of older chaplains regardless of grade, to extend active duty service with other than troop units, such as hospitals, schools and like installations of the Army.
c. Hitherto, chaplains have been subject to the age restrictions in regard to active duty that applied to the officers of all branches. Now, reserve chaplains who are physically fit and under 60 years of age are eligible for assignment to extend active duty in Corps Area Services Command, other than Replacement Centers, but not with troop units.[10]

Arnold extended the maximum age restriction with the understanding that older chaplains would not serve overseas in strenuous physical assignments. Chaplains within this category did not serve in "troop units" but with garrison religious support functions such as chapel duty. Younger and physically fit chaplains remained with troop units. This wartime exception to age maximums and personnel requirements matched capabilities with requirements while meeting personnel number requirements. The establishment of the Army of The United States, Compo 4, allowed for and gave an implied understanding that after the war, older chaplains could apply to the Reserves (Compo 3), National Guard (Compo 2), or return to civilian service.

The Army chaplaincy adjusted personnel requirements in order to meet the wartime emergency. The adjustment of numbers was not a shell game for appearance sake. Adjusting the personnel requirement was a temporary wartime necessity. The War Department and more importantly, the chaplains knew that at the end of hostilities, chaplains' accessioned on this wartime condition would return to civilian life. The exception to allow older chaplains to serve resulted in the chaplaincy meeting the personnel requirement for chaplains in the Army for the wartime emergency.

The Chief of Chaplains continued to maintain the standard for Regular Army commissions. As outlined in Army Regulation 605-30, *Commissioned Officer: Appointments for Chaplains in the Regular Army*, a chaplain must be:

a. A male citizen of the United States.
b. Between the ages of 23 and 34 years.
c. Regularly ordained, duly accredited by and in good standing with some religious denomination or organization which holds an apportionment of chaplain appointments in accordance with the needs of service.
d. A graduate of both 4-year college and 3-year theological seminary courses.
e. Actively engaged in the ministry as the principal occupation in life and be credited with 3 years experience therein.[11]

In contrast, the Navy did not require ministry experience for its applicants. Navy recruitment requirements were significantly smaller than the Army, and the Navy chaplaincy managed to meet its goals during the war. The Army chaplaincy required experience and had significantly more chaplains to recruit. Therefore, Chaplain Arnold continually assessed and temporally adjusted the requirements to meet the wartime requirement.

Engaging with religious denominations and recruitment

Rabbis, priests, and ministers who responded to the call to arms were simply clergymen, as untrained in the ways of the Army as their parishioners, and unskilled in the arts of war. "The Officer Procurement Service delegated full authority for the procuring and processing of Chaplains to the Chief of Chaplains. Completed applications were forwarded by the Chief of Chaplains to the Service for recording and review before transmittal to the Secretary of War's Personnel Board."[12] Unlike World War I, during World War II, chaplains, not civilians nor civilian agencies, retained responsibility and control for religious support to all soldiers.

Early in the war, Chaplain Paul D. Moody, the son of the famous evangelist Dwight L. Moody, had the unique position of being a reserve chaplain and appointed head of the General Commission on Chaplains. Moody, a World War I veteran who served with Bishop Brent, himself a civilian on General Pershing's staff, wanted control over religious support overseas just as Brent had control during World War I. General George Marshall went to Chaplain Arnold with the idea. Chaplain Arnold replied "General, you have a chaplain organization." Chaplain Arnold learned the following lessons from World War I. First, there was no unity of effort. Bishop Brent answered only to General Pershing and in his ministry in Europe placed civilian clergy and auxiliaries on equal footing with Army chaplains. This resulted in some faith groups not receiving religious support. For Chaplain Arnold, the Army and in particular chaplains would oversee all religious support to include auxiliary organizations such as the YMCA and the Knights of Columbus. In addition, unlike World War I, Chaplain Arnold approved all civilian clergy requests to visit troops overseas.[13]

Chaplain Arnold's vision and wisdom to cooperate with religious groups proved invaluable. There were three major endorsing agencies at that time: The Jewish Welfare Board (JWB), The Military Ordinariate that served Roman Catholic applicants, and the General Commission on Army and Navy Chaplains that served most Protestant denominations. As mentioned earlier, Chaplain Arnold brought Chaplain Lev, the first Rabbi to serve in OCCH to interact with the JWB[14] and to assist with chaplain recruitment.[15]

OCCH historically and regularly interacted with the Protestant General Commission on Chaplains. Since every Chief of Chaplains before Chaplain Arnold was Protestant, this relationship between OCCH and the Protestant Commission was well established.

Chaplain Arnold coordinated with his own denomination and was able to influence Bishop John O'Hara of the Military Ordinariate to publish circular letters similar to OCCH circular letters as early as January 1941. The Military Ordinariate published 44 circular letters by June 1945. They provided excellent publicity to Roman Catholic parishioners, and wise and authorized advice to Roman Catholic chaplains. For example, in Circular Letter 33 Military Ordinariate, August 14, 1943, "They [Catholic Chaplains, ed.] were informed by "His Excellency the Apostolic Delegate that the Holy Office….to allow chaplains located in places where it is difficult to obtain wine, permission to use only water in the two ablutions following Holy Communion."[16] These Military Ordinariate circular letters achieved the same impact as the Chief of Chaplains circular letters by connecting Roman Catholic priests to their denomination and one another when they performed their sacred duties in the Army Chaplaincy.

Chaplain Arnold wanted every chaplain to assist with recruitment. Circular Letter 265, January 2, 1943, directed all Chaplains to assist with the procurement of chaplains in order to double their numbers in one year. Publicity, encouragement, and teamwork motivated the chaplains to reach their procurement goal for 1943, 1944, and 1945.[17]

At the start of the war, only male citizens could serve as chaplains. Several groups challenged this rule. An organization of women ministers and other individuals believed the training school for the newly established

Women's Army Auxiliary Corps, and any large organization with female soldiers, should have female chaplains. A survey at the training camps showed that very few of the women desired female chaplains. As a result of this survey, the War Department refrained from authorizing female chaplains. Many female ministers did become chaplain's assistants or directors of religious education. When the need for chaplains became acute, the citizenship rule changed, and the appointment of citizens of cobelligerent and friendly powers could join the chaplaincy, but this did not extend to allowing women.[18] A chaplain of the Free French and three chaplains from the Philippine Army attended the Army Chaplain School.[19]

A sign of leadership is its influence beyond the chain of command.[20] Chaplain Arnold's ability to interact and influence across denominational lines shows confidence yet a willingness to understand others.

Managing Personnel Deploying Overseas

The need for chaplains to deploy overseas was a challenge from the very beginning of the war. Just a few days after the bombing of Pearl Harbor, the Army published Army Regulation 605-10, *Commissioned Officers: Officers Appointed in the Army of the United States* exempting chaplains and medical personnel from the required six-month waiting period before deploying overseas.

> Persons appointed from civilian status, except for the Medical Department or for duty as a chaplain, will not be assigned to duty with units of the field forces unless they have completed not less than six month active military serve as a commissioned officer subsequent to initial appointment, and have satisfactory completed an appropriate course of

> instruction at a special service school of the army or service to which they are assigned.[21]

Despite this regulation, Chaplain Arnold, as the proponent and assignment authority for Army Chaplains, did all he could to make sure prepared and qualified chaplains deployed overseas to the critical ministry of providing religious support.

However, an issue arose when on February 3, 1943 a German submarine torpedoed the U.S.A.T. Dorchester, a transport ship sailing from New York to England. Four chaplains died that day. They gave their life jackets to other soldiers. The story of two Protestants, one Catholic, and one Jewish Rabbi became a symbol of unity and sacrifice. It appears that one of the chaplains, Chaplain Clark Poling, did not attend the chaplain school.[22] Some chaplains did go overseas with no military education. Typically, these chaplains had prior service or went to non-hostile locations.[23]

Chaplain Arnold and his staff must have been frustrated that regardless of their efforts, chaplains deployed without the proper training.[24] Just over two weeks after the sinking of the Dorchester, Chaplain Arnold wrote the following, on February 15, 1943.

> Volunteers for Foreign Service.—From time to time this office must name chaplains for oversea duty. In order that this office may have a list of volunteer chaplains for this duty, it is suggested that chaplains desiring overseas service submit a request through channels. Only those chaplains with the following qualifications should apply for foreign service: under 45 years of age; available unlimited service; and more than six months experience in the Corps of Chaplains. Since approval of such transfers must be obtained before action can be taken, request

should not be submitted direct to this office but through Commanding Officers concerned.[25]

Chaplains could volunteer for "Foreign Service." However, OCCH wanted to remind chaplains all requests were subject to Chaplain Arnold's approval. All requests had to go through Command channels; keeping religious support planning and operating within command channels is essential for the unity of effort. When the need for chaplains became critical on the continent of Europe, Chaplain Arnold deferred to the commands, and without exception to policy, sent chaplains directly from the Chaplain School to overseas duty.[26]

Procurement and the quota system

Selecting the right civilian clergy to fulfill the duty of providing for the free exercise of religion for all Army soldiers presented unique challenges. In addition to the requirement of one chaplain for every 1,200 soldiers, the additional requirement of providing for the diverse religious denominations presented a challenge to the Personnel section. After World War I, the Army addressed this issue by establishing a quota system based upon the national census. Each denomination received a specified number of authorizations in the Regular Army and the Reserves based on the national civilian population.[27]

The procurement goal was different from the quota goal given to each denomination. During World War II, the Army chaplaincy met the procurement goal of the number of chaplains needed in the Army overall. However, the chaplaincy did not meet the denominational quota established at that time.[28]

In August 1946, the War Department published War Department Pamphlet 16-1, *The United States Army*

Chaplaincy. The document states that it is not "the official history of the Corps of Chaplains" but provided as "background and source material" for Army Chaplains.[29] The author Chaplain Roy Honeywell wrote that one of the major problems during World War II was the quota system.

> In spite of all the preparedness, however, the facilities of the Office of the Chief of Chaplains and of many denominational and interdenominational agencies were severely taxed in an attempt to secure the right type of men for the thousands of chaplain positions which had to be filled. Although fairness was maintained by a revised quota system, these quotas complicated the work of procurement.[30]

During the war, some denominations aggressively sought to fill their quotas while other denominations did not. When it became apparent that some denominations would not meet their quota, authority to allow 135 percent over a denominations quota allowed the chaplaincy to meet its procurement goal.[31] However, with the postwar drawdown, these denominations had to cut back the number of chaplains authorized, while those denominations who did not meet their quota did not have to reduce their number of authorizations in the Regular Army.

The quota system was also problematic regarding African American Chaplains. Four of the major African American denominations at the time, African American Episcopal, African American Episcopal Zion, Baptist-Colored, and the Colored Methodist Episcopal[32] did not meet their quota goal of 790 chaplains. Only 185 African American chaplains were on duty on September 2, 1945,[33] thus only reaching 23 percent of their quota goal.[34] However, several African American chaplains did serve from several other denominations to include the first African American Roman Catholic Priest to serve in the Army, and therefore, the classification of "black," and "white" denominations was problematic and perpetuated racial biases.[35]

African American Chaplains and lay leaders

There were only three African American active duty chaplains in 1934.[36] Robert Gushwa points out that the unintentional education requirements provided the greatest obstacle to recruiting African American Chaplains:

> [African American, ed.] churches were authorized 790 chaplains but only 174 were in service at the end of the war. While many [African American, ed.] chaplains came to the Army from churches that were not listed as " [African American, ed.]," the fact remains that only 22 percent of the quota was filled. Although it was the policy to assign [African American, ed.] chaplains to [African American, ed.] troops, obviously this could not always be done. The educational requirements for [African American, ed.] ministers were modified in such a way that many ministers who were barred by previous regulations were able to qualify.[37]

The chaplaincy wanted African American chaplains and established the same authorizations regardless of race. However, the desire to have educationally trained professionals automatically disqualified a significant portion of African Americans, since clergy within many African American churches did not require formal education. OCCH adjusted the education requirements in an attempt to meet quota and accommodate for those faith groups.

Figure 21. African American Lay Leader conducts Protestant worship service. New Guinea, 1944. Image from the author.

The use of African American lay leaders filled a significant gap in providing religious support in the segregated units. The After Action Report for the Pacific Theater provides several examples of lay leaders who provided supervised religious services:

> Many of these men were ordained before coming into the service...Every [African American, ed.] unit, with one exception, of this Command has one or more preachers. They conduct Sunday and Weekday Services, an undertaking rarely found in white units...The Services are dignified, devotional, and moving.[38]

Under the authority of the commander, chaplains supervised and supported these lay leaders.[39] Although there was a shortage of African American chaplains during the war, OCCH adjusted the requirements and utilized lay leaders to provide religious support. For comparative purposes, chaplain supervision for current chapel lay leaders and contractors who serve in the chapel, and chapel auxiliaries such as the Protestant Women of the Chapel and Catholic Women of the Chapel, requires monitoring at every echelon as they provide religious support to soldiers and families.

The chaplaincy led the way in addressing racial issues of the time. All chaplains regardless of race or religion trained together at the chaplain school.[40] Circular Letter 262, November 2, 1942, included a handbook by Edwin R. Embree, the grandson of the well-known abolitionist John Gregg Fee. Embree lists notable African-American authors and addresses the "lack of respect and treatment by whites."[41] Circular Number 268, February 15, 1943, mentions the visit to OCCH of an African American Chaplain who had served in a segregated unit on the ALCAN (Alaska-Canada) highway. Chaplain Arnold highlighted the chaplain and his ability "to develop recreation programs and show men how to be good in their everyday life, not just to preach many sermons and hold church services on Sundays."[42] Chaplain Arnold addressed abuses too. In Circular Number 281, February 1, 1944, Chaplain Arnold wrote:

> Chaplains are advised to use the word "[African American, ed.]" when referring to a person belonging to the black race. Reports have reached this office indicating that some chaplains have used other terms in ordinary speech and public address. This is unnecessary and a hindrance to good relationship between members of our American Democracy. It is believed that chaplains should do everything in their power, by good example, to

> avert bitter feeling and strife on the part of racial groups serving in the armed forces.[43]

Another example of the Army chaplaincy progressing in the cause for racial equality is the *Song and Service Book for Ship and Field*. This hymnal contained several "Folk Songs" in the back that reflected racial stereotypes of the times.[44] The After Action Report for the European Theater of Operations mentioned the Folk Songs "were hardly used" by chaplains and the key of the hymnal made it difficult for servicemen to sing.[45] On May 14, 1945, Chaplain Assistant Private Charles W. Warren, under the guidance of his chaplain, Chaplain Francis W. Fero transcribed the 147 hymns into a key reachable for servicemen.[46] They left out the insensitive "Folk Songs" in the back.[47]

The Army chaplaincy adhered to religious freedom, which was the bedrock of the United States. Although not perfect, Army chaplains and chaplain assistants led by Chaplain Arnold worked through the biases, prejudices, and advanced the cause of liberty and freedom by treating all people with dignity and respect.

Chaplain Assistants

There was no official enlisted branch within the Army Chaplaincy during World War II. The Army Organization Act of 1950 started the process of formalizing a structure for the chaplain assistant. World War II had highlighted the need within the chaplaincy for someone to assist the chaplain in the performance of his duties.

General Orders 253, December 28, 1909, authorized one enlisted soldier to assist the chaplain in the performance of his duties. The only qualification was a high moral character. The assistant was not to be a personal attendant to the chaplain. Selection of this soldier was usually an agreement between the chaplain and the commander. When no suitable soldier could serve as a chaplain assistant, the chaplain would work by himself and recruit soldiers to play the organ for worship services. After the National Defense Act of 1920, the first official Army Regulation for chaplains, Army Regulation 60-5 *Chaplains, General Provisions*, echoed what was already in Army Regulation 210 70, *Commanding Officer*. The duty of the Commander was to "detail such needed assistants to chaplains as may be deemed desirable and practicable by commanding officers."[48]

During the interwar years, there were several cases when the chaplain did not receive an assistant from the command. Monthly circulars addressed this issue with the "hope of finding a satisfactory solution."[49] One circular letter implied the amount of work a chaplain performs would influence the command to provide a chaplain assistant. The following is an example of a justified needs statement for manpower analysis for Force Management.

> A COMPETENT ASSISTANT was secured by a chaplain, and how? This chaplain, shortly after reporting for duty, went to his commanding officer with a proposed outline of character building activities, which included such headings as "Religious services", "Midweek classes for instruction both religious and secular", "Sunday School", "Programs of entertainment", "Chaplain's Bulletin, a mimeographed weekly paper", "Post Library", "Glee Club", "Counseling", "Pastoral work", "Civilian contacts", "Sing songs", "Men's brotherhood", "Instructor for over fifty students in the Chaplains' Extension Course", "Summer training camps", "CCC work", etc., etc. A brief conference followed. The commanding officer, impressed with the earnestness and efficiency

> of his chaplain and the comprehensiveness of his program, saw at once that his chaplain had more than a one man job and that he needed an assistant having special educational and character qualifications; and the way was opened for him to secure competent help.[50]

Throughout the interwar years and during World War II, finding qualified assistants presented a challenge for the chaplaincy. Chaplain Assistants are combatants and therefore are not under the Geneva Convention as non-combatants. Chaplain Assistants provide security in hostile areas for their chaplains. The requirement for security within the continental United States was not as great; therefore, Chaplain Arnold relied on Army Regulation 60-5, *Chaplains*[51] guidance to coordinate with civilian secretaries to serve as assistants within chapels to perform the administrative tasks.[52] Similar to the policy of allowing older chaplains to serve in garrisons in the United States, Arnold applied the same principle of matching capabilities with requirements by allowing civilians to serve as administrative assistants. The statement of General George Marshall is of interest in this connection.

> The thought has just occurred to me that the influence of the chaplains might be amplified by giving them temporarily several really qualified assistants. Since the fighting will be finished in Europe, the manpower problem will not present quite the usual obstacle. It may be a good idea, worth looking into. A chaplain might use a half-dozen picked men with profit.[53]

Finding qualified assistants presented a problem. The assistant was one who was skilled in many tasks, not just music. As one Pacific Theater combat chaplain wrote, "Having a qualified assistant can relieve him of endless details which deprive him of time to spend with the troops."[54] He had to carry the Estey pump organ, field desk, hymnal chest, chaplain's portable altar, files, personal gear and things for soldiers like books and games. The amount of equipment for the chaplain team led one assistant to ask his chaplain, "Sir, didn't Jesus travel lighter than this?"[55]

One chaplain wrote Chaplain Arnold to "express my appreciation for the splendid cooperation of the personnel officer in selecting a good assistant for the Chaplain's work and the splendid spirit that was shown by the company commander in releasing men to be tried out for the job."[56] However, when a chaplain found a good assistant, he or she was stuck at the rank of Technical Fifth Grade 5 (Current Grade of E-4). Those assistants who performed their duties well and wanted to attain higher rank, transferred to another branch specialty.[57]

Picking the right assistant for the mission is very important. Although the following example is that of a segregated unit, the purpose and function remain the same regardless of the unit. The After Action Report for the European Theater of Operations found that placing an African American assistant to the chaplain was "decisively" advantageous for the following three reasons. First, he can provide better liaison between the chaplain and the enlisted person. Second, he can eat and live with the enlisted men of the command "without embarrassment." Finally, an opportunity for a qualified soldier to hold a position that "confers a certain measure of prestige among his fellows."[58]

It appears that many times, the chaplains did not always receive the best soldiers to perform their sacred duties. World War II veteran Kurt Vonnegut highlighted the reputation of assistants in the post-war semiautobiographical *Slaughter*

House Five. "The chaplain assistant cut a pathetic figure in the U.S. Army. Nothing but an object of scorn to his comrades, he was incapable of inflicting harm on the enemy or of providing support for his fellow soldiers."[59] Comments from After Action Reports from the European and Pacific Theater stressed the struggle with receiving unqualified assistants from the command and the importance of having the right assistant in order to provide religious support.[60] Sometimes when a chaplain changed an assignment, the new chaplain would not always find the current assistant in his new assignment satisfactory. Some chaplains when transferred would take their assistant with them. The limitation in rank structure for chaplain assistants led chaplains in both the European Theater and Pacific Theater to arrange with higher headquarters to allow their assistant to transfer with them to their next assignment.[61]

World War II provided the first institutional training for Army chaplain assistants. In June 1944, the Army Air Force opened a two-week course for their chaplains with a parallel course for enlisted men. Chaplain Carpenter, the Senior Air Chaplain, would meet every chaplain who would serve in air units for "orienting them to their jobs with air units."[62] Chaplain Carpenter also oversaw the selection of the assistants. An Air Force order directed that Chaplain Carpenter must first declare all applicants for the chaplain school as surplus personnel within the Air Force. The order allowed Carpenter to screen all personnel in order that they meet his emphasis upon the mission requirements, which was musical ability. The course trained assistants for choir directing, automobile operation, office procedures, and the unique challenges within the Air Force. All but six of the 945 airmen completed the course of the twenty sessions. Seventy-four were members of the Women's Army Corps.[63] Of the 945 assistants that attended the Chaplain School, 312 were organists, 412 pianists, 570 choristers, and 155 vocal artists.[64]

In the autumn of 1944, OCCH noted the success of the Air Force Chaplain School, and developed and recommended a three-week course of instruction for assistants at the Army Chaplain School. However, due to his inexperience Chaplain Luther Miller, the new Chief of Chaplains, was unable to navigate the bureaucracy within the Pentagon and the Army Chaplain Assistant course never started during World War II. Then in the spring of 1948, at a conference for training officers, chaplains recommended only on the job training for the assistant. These chaplains forgot the strong recommendations from the After Action Report from the European Theater that showed a marked difference between those "who graduated of the Army Air Force training course for chaplains' assistants in the United States and those assistants who, recruited in Europe, had not had the benefit of such training."[65] Not until the dawn of the Korean War on February 7, 1950, did the first official Army chaplain assistant course begin for the Army. This four-week course included 67 men and two women. Meanwhile, the Air Force school for chaplain assistants did not stop training except for a brief period in 1950 when the Air Force needed the classroom space to train new chaplains.[66] Within a year, the Air Force school opened again to train new assistants for the Air Force.[67]

Chaplain Arnold's ability to understand personnel systems and adjust requirements to meet the wartime emergency meant that trained and certified chaplains, and for the Army Air Force, chaplain assistants, provided adequate religious support. Chaplain Arnold and OCCH facilitated and encouraged the bond with the chaplain's denomination that resulted in trust and transparency between the chaplain, the chaplain's denomination and OCCH. Chaplains and chaplain assistants had pride in their calling and service that strengthened their resiliency and identity. The result was a nation joined with the Army into a common purpose of religious freedom and toleration.

Notes

[1] U.S. Department of the Army, Army Regulation 5-22, *The Army Force Modernization Proponent System*, (Washington, DC: U.S. Department of the Army, October 28, 2015), 13.

[2] Michael Snape, *God and Uncle Sam: Religion and America's Armed Forces during World War II, (*Boydell Press, 2015), 5.

[3] Robert L. Gushwa, *The Best And Worst of Times: The United States Army Chaplaincy 1920-1945*, (Volume IV, Office of the Chief of Chaplains, Department of the Army, 1977), 86.

[4] Component (Compo) 1 is Regular Army. Compo 2 is National Guard. Compo 3 is Reserves. Compo 4 is the Army of the United States.

[5] Gushwa, 96.

[6] Ibid., 99.

[7] David B. Jorgensen, *Air Force Chaplains, 1947-1960*, (Volume II, Office, Chief of Air Force, 1961), 89.

[8] Ibid.

[9] During WWII General Marshall kept a picture of WWI soldiers worshiping in a bombed out cathedral in France. Marshall had an artist paint this picture to hang in his office for the remaining of the war.

[10] Memorandum to All Chaplains Subject Interpretation of the War Department policy relaxing the age restrictions for extended active duty for chaplains. April 7, 1941.

[11] U.S. War Department, Army Regulation 605-30 *Commissioned Officer: Appointments for Chaplains in the Regular Army* (War Department, Washington D.C.,

December 16, 1941), 2.

[12] Headquarters, Army Service Force: *Officer Procurement During World War II*. Army Service Forces. Archives Section, Fort Leavenworth, Kansas, 73, http://cgsc.contentdm.oclc.org/cdm/singleitem/collection/p4013coll8/id/4443/rec/1 (accessed December 18, 2017).

[13] Gushwa, 106-107.

[14] Roy J Honeywell, Chaplains of the United States Army, (Washington, DC: Office, Chief of Chaplains, 1958), 237.

[15] Headquarters Army Service Forces. Circular 212, Office of the Chief of Chaplains, December 2, 1940.

[16] Military Ordinariate, Circular Letter 33, (462 Madison Ave NY, NY, August 14, 1943).

[17] Army Service Forces, *The Chaplain Serves: Chaplain Activities, 1943*, (Washington D.C. Office of the Chief of Chaplains 1 March 1944), 6.

[18] Gushwa, 101.

[19] Ibid., 110.

[20] U.S. Department of the Army, Army Regulation 600-100 *Army Profession and Leadership Policy*, (Headquarters, Department of the Army, Washington D.C. April 5, 2017), 1-11.

[21] U.S. War Department, *Commissioned Officers: Officers Appointed in the Army of the United States*, Army Regulation 605-10 (War Department, December 10, 1941) 13.c.

[22] In Circular 252, June 1, 1942 Chaplain Arnold directed all personnel assigned overseas to submit to his office the name and address for their next of kin. Chaplain Clark Poling father, Daniel A. Poling regularly interacted with the Chief of Chaplains Office and therefore, Chaplain Arnold must have provided pastoral care to Daniel Poling

when he returned from his trip to London.

[23] Those chaplains that did not attend the chaplain school received advanced and specialized training after the war.

[24] Chaplain Roy Honeywell, the OCCH Operations Chaplain during war never mentioned the Four Chaplains in the first Official History of the Army Chaplaincy. Chaplain Daniel Jorgensen, the author of the Army Air Force history, 1917-1945 only provides four sentences to the "ill-fated" Dorchester.

[25] Circular 268, Office of the Chief of Chaplains, February 15, 1943.

[26] *Report On The Army Chaplain In The European Theater* (The General Board, United States Forces, European Theater, Study Number 68), Chapter 2.

[27] Honeywell, 214-215.

[28] Ibid., 215-221. Provides insight to the quota system during WWII.

[29] U.S. War Department, War Department Pamphlet 16-1, *The United States Army Chaplaincy*. (War Department, August 1946), iii.

[30] Ibid., 21.

[31] Honeywell, 215.

[32] Original titles of denominations are kept for historical purposes.

[33] The reason for the difference in the numbers between Honeywell and Gushwa is that Honeywell counted the 11 African American chaplains who served in denominations not categorized as African American.

[34] Honeywell, 215-216.

[35] For further study about African American chaplains and religious liberty in the Army read Chaplain (MAJ) John Pearson. *The Black Experience in the Military Chapel*,

DA PAM 16-104 Military Chaplain's Review (Dept of the Army, Winter, 1975). For more insight into the problems of the quota system and race read Chaplain (LTC-P) Jerry L. Robinson. *A Chronological Record of Historical Events Relating To Diversity In The U.S. Army Chaplaincy As Viewed By Chaplain (Major General) (Retired) Matthew A. Zimmerman, Jr*, Strategy Research Project, (U.S. Army War College, Carlisle Barracks, PA, 1997).

[36] Gushwa., 76.

[37] Ibid., 100.

[38] Army Forces Pacific, *History of Chaplains' Activities in the Pacific*, (Chaplain Section, Ground Headquarters, Army Forces Pacific, 1946), 400.

[39] Chaplain Arnold learned from WWI the importance of overseeing all lay leaders and auxiliaries. During WWI, organizations such as the Young Men's Christian Association (YMCA) and Knights of Columbus operated independently within the Army and on occasion conflicted with the Army mission. Chaplain Arnold placed Chaplains over these organizations and lay leaders to ensure synchronized ministry within the Army mission and the Chaplaincy vision.

[40] Jorgensen, *The Service of Chaplains to Army Air Units, 1917-46*, 282-283.

[41] Circular Letter 262, Office of the Chief of Chaplains, November 2, 1942.

[42] Circular Letter 268, Office of the Chief of Chaplains, February 15, 1943.

[43] Circular Letter 281, Office of the Chief of Chaplains, February 1, 1944.

[44] Evan L. Bennett, ed., *Song and Service Book For Ship and Field.* (U.S. Government Printing Office Washington, 1942), 183.

[45] *Report on the Army Chaplain in the European Theater*, 103.

[46] Ibid., 112.

[47] Francis W. Fero and Charles W. Warren, eds., *Song and Service Book for Ship and Field with Music Transposed.* Unpublished. Original with Chaplain (Colonel) Robert Nay.

[48] U.S. War Department, Army Regulation 60-5, *Chaplains*, (Washington, DC. War Department, February 15, 1924), para 10.

[49] Circular Letter 96, Office of the Chief of Chaplains, May 10, 1932.

[50] Circular Letter 164, Office of the Chief of Chaplains, July 8, 1936.

[51] U.S. War Department, *Chaplains*, Army Regulation 60-5, *Chaplains*, (War Department, February 20, 1941), para 4.

[52] David B. Jorgensen, *Air Force Chaplains, 1947-1960*, (Volume II, Office, Chief of Air Force, 1961), 125.

[53] *Report On The Army Chaplain In The European Theater*, 32-33.

[54] *History of Chaplains' Activities in the Pacific*, 193.

[55] Christopher Cross and William R. Arnold. *Soldiers of God* (New York: E. P. Dutton, 1945), 86.

[56] Circular Letter 220, Office of the Chief of Chaplains, March 15, 1941.

[57] *Report On The Army Chaplain In The European Theater*, 28.

[58] Ibid., 29.

[59] Snape, 187.

[60] Headquarters, U.S. Army Troops, Office of the Chaplain. CONFIDENTIAL *Report Activities of Chaplains, 7th Infantry Division, Battle of Attu*, (Headquarters, U.S. Army Troops, Office of the Chaplain, July 5, 1942), 17, 20.

[61] *Report On The Army Chaplain In The European Theater*, 28. *History of Chaplains' Activities in the Pacific*, 80.

[62] Office of the Air Force Chaplains, *A History of the Air Force Chaplain*, (Extension Course Institute, Air University, Litho Press Inc., San Antonio, Texas, May 1966), 18.

[63] Honeywell, 250-251.

[64] Honeywell, 251.

[65] *ETOUSA Special & Morale Services Guide* (Special Service Division, ETOUSA, May 1944) 50. *Report On The Army Chaplain In The European Theater*, 28.

[66] Honeywell, 318.

[67] Jorgensen, *Air Force Chaplains, 1947-1960*, 119.

Chapter 8

Facilities

Facilities play a crucial part in the nation's defense. The Army defines facilities as "Real property consisting of one or more of the following: a building, a structure, a utility system, pavement, and underlying land."[1] A suggested religious support definition for facilities is "Real property such as buildings that supports the free exercise of all religious faith groups in the Army." The greatest legacy of Chaplain Arnold's leadership is the World War II chapel that still stands on many military installations. Providing houses of worship for all faith groups remains the enduring foundation for the very purpose for Army chaplains throughout all generations.

Facilities include training facilities, office, and storage space, and living quarters. Facilities are expensive to build and maintain. Facilities for chaplains within the DOTMLPF-P analytical framework must consider the similar requirements for other special staff branches within the Army. Besides common office space, lawyers require courtrooms and doctors require hospitals and clinics. Although in tactical combat settings, makeshift locations and rooms do provide temporary function, the established and permanent facility provides the essential requirements for the capability of each special staff officer. For the chaplain, the chapel is essential for providing religious support.

The Army Chapel

Chaplain George J. Waring's *Chaplain's Duties and How Best to Accomplish His Work*, published by the War Department just before World War I, highlight the importance of the chapel.

> To my mind there is nothing that helps the chaplain's work as much as the presence in a post of a chapel [sic]. But the chapel ought to be no mere compromise, that is, a combination of school, library, recreation hall, and chapel. It should

be a distinct building of distinctly ecclesiastical architecture, and it should be erected in that part of the post where it will be most easily accessible to all persons living theron [sic]. If more religion could be instilled into the lives of our soldiers there would be less need for hospitals and guardhouses; discipline would be more easily maintained; contentment would be far more general. It is a wonderful thing what an influence for good religion is among soldiers. They want to be religious and they are desirous of attending service, but they usually get so little encouragement, not alone from one another, but also from the lack of proper facilities and sufficient opportunities, and not the least from the fact that even the chaplains are often too much taken up with worldly pursuits, e. g., canteens, bakeries, etc., that the poor fellows frequently give up religion as an impossibility. And yet I am thoroughly convinced that many men who now stay away from church would go if they could attend in a building which would exude a religious atmosphere, and which would be entirely dissociated from secular pursuits and enjoyments. Such a chapel would, by its very presence there, be a constant though silent reminder to the men that there is a building set aside, as there is a day set aside, for the worship of Almighty God. This would help greatly, but more is needed.[2]

Funding did not occur for chapels during World War I nor in the post-war drawdown. Monetary gifts exceeding $100,000 helped build several chapels, such as the small chapel at Arlington Cemetery and Memorial Chapel at Walter Reed General Hospital.[3] During the Depression era, The Works Projects Administration allowed for the building of nine chapels utilizing the Civilian Conservation Corps as labor. However, these chapels resembled the local architecture and usually accommodated the local religious traditions of the area. The desire of soldiers for a house of worship led to several unusual initiatives despite the lack of a systematic effort and official resources. In one case, without the help of the government, resourceful soldiers stationed at Will Rogers Field, Oklahoma City, used the wood from an old chicken coop and a corn crib to build a house of worship that seated one hundred soldiers.[4]

Despite the lack of uniformity of facilitates for all religious faith groups during the 1930s, Chaplain Alva Brasted's last words as Chief of Chaplains in 1937 touched on the impact of these chapels.

> The nine new Army chapels built by the Government in the last few years have advanced the moral and religious work of the military service very materially. These chapels have helped morale, have caused a very decided increase in religious attendance, have contributed to the prestige and influence of the chaplain's work, have increased the interest of civilian communities in the work of the chaplains, and have brought civilians and Army personnel closer together in acquaintance and fellowship, and in appreciation of each other…Nothing has been done by our Government for the military service that has met with more wholehearted approval on the part of good citizens than the building of Army chapels. It is our most earnest prayer that the good Lord may open the way so that this great need for many more Army chapels shall be supplied.

Figure 22. Early Drawing of World War II Cantonment Chapel.
Source: Arlington Chapel dedication memento.

In 1940, only seventeen Army posts had permanent chapels.[5] By September 2, 1945, V-J Day, there were 1,532 Army chapels in use in the United States. [6] Chaplain Arnold persevered through the bureaucratic system and gained the trust and support of the leadership in the War Department to resource this significant effort. On March 17, 1941, President Roosevelt signed congressional bill HR - 361 7, Public 13, authorizing a master plan to build 604 chapels for $12,816,880 within six months. March 20, 1941, War Department Bureau of Public Relations released the following statement, "604 Chapels will be built throughout the Army…Where soldiers of the Protestant, Catholic and Jewish faiths may make their devotions in an appropriate settings."[7] Additional appropriations from Congress built more chapels throughout the war.

Chaplain Arnold personally oversaw the original design of the chapel and the construction process to make sure the facilities accommodated all faith groups within the allocated budget. Unlike the Works Project chapels that exceeded the cost of many of the local religious buildings, the new Army chapel was going to be modest. To keep costs down, the materials for the chapel were only to last a generation. Chaplain Arnold also oversaw the building process. Contractors were substituting cheaper material and material that did not meet the standard and intent for the chapel. For example, some contractors were using clear glass instead of cathedral glass. Chaplain Arnold utilized the monthly circular letters and encouraged all chaplains to "become familiar with the specifications of the chapel in his area and secure permission from his Commanding Officer to make frequent inspections to see that the contractor conforms to the War Department specifications."[8] The Quartermaster and Construction officers who were veterans from the Depression era Works Project Administration had a working relationship with Chaplain Arnold and OCCH. Early in the process, Chaplain Arnold wrote, "We have already headed off some monstrosities."[9]

Figure 23. World War II Cantonment Chapel with the Steeple on the side
Source: The author's personal collection.

However, one chapel at the Orlando Army Air Field required special attention. The designated place for the chapel mistakenly was too close to the runway. The chapel was already under construction, when it was discovered that the steeple interfered with the approach and landing for the aircraft. To conform to the contract, and the intent of the design of the chapel, the local leadership compromised and built the steeple on the ground, alongside the chapel. A service member sent Chaplain Arnold a postcard of this chapel. This must have meant something to Chaplain Arnold since he kept it and included it with his papers that are currently located at the George C. Marshall Library.

Chaplain Arnold had to contend with complaints within his denomination that the chapels were not large enough and did not have a confessional for the Roman Catholic soldiers.[10] Arnold's response was to build more chapels and provide more services. For the confessional, Arnold recommended the "folding priedieu type…since some men are deterred from confession if they have not the protection of the screen."[11]

The Army chapel is symbolic for the Army chaplaincy. Therefore, the first multi-faith Army Chapel required a symbolic display for the public. On July 27, 1941, General George Marshall, Major General Edmund Gregory the Quartermaster General, and Chaplain Arnold led a ribbon-cutting ceremony for the first Cantonment Chapel at Arlington, Virginia. General Marshal praised Chaplain Arnold for his "care for the moral and spiritual guidance of its soldiers." Marshall's intent for these chapels was good citizenship and to "have a clean army, morally and physically, and these chapels are very important contributions to that end." Major General Gregory, the Quartermaster General echoed the freedom these chapels represented. "There is nothing in construction that could stamp it as so distinctively American…because only in a free country could you find a church built to be used for worship by Catholic, Protestant and Jew alike." Chaplain Maurice W. Reynolds, a Protestant, provided the invocation and echoed the same thoughts. "Grant that in the days to come this Thy Tabernacle may be a source of comfort, of strength of character, of courage, and of hope to the soldiers who shall here serve their country and their God." Chaplain Arnold's words reflect what he believed the power these chapels had for the fighting force. "To every man who enters this House of Prayer God says 'Without Me you can do nothing'; to every man who comes forth from this House of Prayer God has said 'With Me you can do all things'."[12]

People who attended the ceremony and helped make the building of the chapels possible received a memento scrapbook with pictures, bulletin, speeches of all key participants, as well as a piece of the ribbon from the ribbon cutting ceremony.[13] The monthly Circular Letter 244, February 1, 1942, contained a piece of the ribbon taped to the letter. A parachute infantry commander who wrote a letter to Chaplain Arnold reflected the importance of the impact these chapels had upon soldiers.

> For the first time in over nine years since leaving West Point I worshiped in a chapel (one of the new type) on an Army Post that from a material side was satisfying to the inner man....It is however very pleasing and satisfying to see at long last chapels appearing on the landscape of our Army posts and camps. I know you didn't personally build these chapels but you are the driving force behind it all. Please accept from a junior sincere congratulations in leading our chaplains to a much higher standard of service and common good than I observed when I first came into the service.[14]

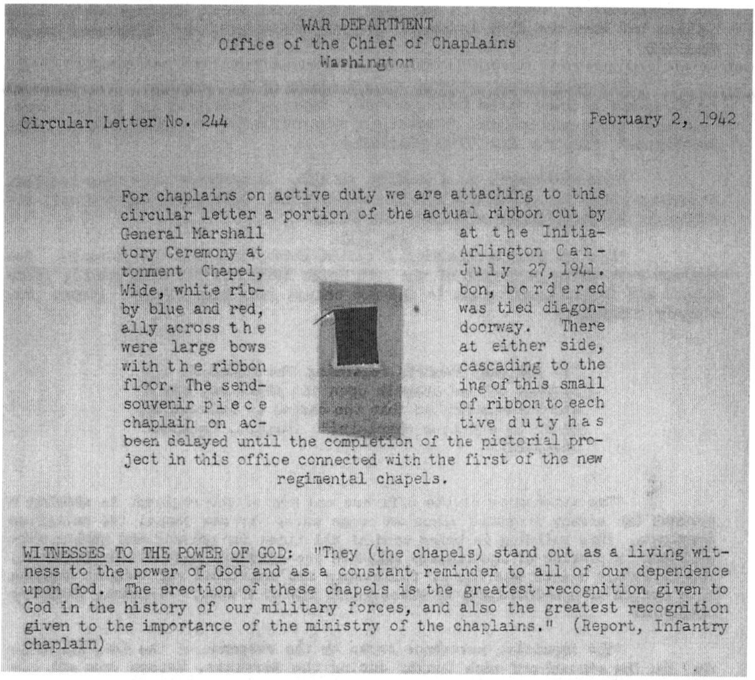

WAR DEPARTMENT
Office of the Chief of Chaplains
Washington

Circular Letter No. 244 February 2, 1942

For chaplains on active duty we are attaching to this circular letter a portion of the actual ribbon cut by General Marshall at the Initiatory Ceremony at Arlington Cantonment Chapel, July 27, 1941. Wide, white ribbon, bordered by blue and red, was tied diagonally across the doorway. There were large bows at either side, with the ribbon cascading to the floor. The sending of this small souvenir piece of ribbon to each chaplain on active duty has been delayed until the completion of the pictorial project in this office connected with the first of the new regimental chapels.

WITNESSES TO THE POWER OF GOD: "They (the chapels) stand out as a living witness to the power of God and as a constant reminder to all of our dependence upon God. The erection of these chapels is the greatest recognition given to God in the history of our military forces, and also the greatest recognition given to the importance of the ministry of the chaplains." (Report, Infantry chaplain)

Figure 24. Monthly Circular Letter with Ribbon from Ceremony
Source: The author's personal collection.

Chaplain Arnold's influence in building chapels extended beyond the Army. The construction of over 600 Army chapels and the public display of support and enthusiasm of the dedication of the chapel at Arlington woke the Navy from its slumber. Two months after the Arlington dedication ceremony, the Navy actively pursued funding for new chapels. The definitive chaplain history book published by the Navy states, "Records show that distressingly little was done to provide adequate chapels for naval personnel before Pearl Harbor."[15] The Navy pursued funding for its chapels based upon Arnold's chapel design.

Even outside of the military, many civilian churches throughout the United States displayed collapsible cardboard miniature replicas of the Cantonment Chapel as reminders of the religious support for their loved ones in the Service.[16] At traveling recruitment and bond drives, a manufactured frontage of the chapel including the entrance and the steeple provided a powerful visual example of how the Army provided worship for Army soldiers. The civilian community would enter through the front into a tent with numerous displays of chaplain equipment, and literature informing the population of military efforts to protect the free exercise of religion.[17]

Finally, Chaplain Arnold's purest and enduring legacy is in the identity and the naming of the Army chapel. In Circular Letter 272, May 1, 1943, Chaplain Arnold wrote:

> Our chapels have been built by a government which declares that man shall be free to worship God as seems best to himself. They have been so designed that men of all creeds and faiths may enter therein and worship. We have a letter from the Secretary of War in which he states that the War Department looks with disfavor on the designation of these chapels as memorials to individuals or by names having a particular denomination connotation. The Chief of Chaplains believes that chapels should be designated by numbers (1, 2, 3) or by letters (a, b, c) or preferably by regiments or areas for which the chapel was built.[18]

The chapel represents the soldier's freedom to worship. The object is the soldier, not the chaplain or the faith. A chaplain's identity remains faithful to their faith while serving and protecting the faith of all Soldiers. As General Marshall said, providing the facilities for soldiers to practice their faith is essential for a "clean army, morally and physically."[19] Army chapels continue to this day to be a reminder and a provision for religious freedom and toleration.

Notes

[1] U.S. Department of the Army, Army Regulation 5-22, *The Army Force Modernization Proponent System*, (Washington, DC: U.S. Department of the Army, October 28, 2015), 13.

[2] George J Waring, *Chaplain's Duties and How Best to Accomplish His Work*. (Washington, DC: GPO, 1912). Page 3-4.

[3] Headquarters Army Service Forces. Circular Letter 55, Office of the Chief of Chaplains, May 27, 1930.

[4] War Service of the National Lutheran Council: 1940-1948. *By Their Side: A Memorial* (Bureau of service to Military Personnel National Lutheran Council, 1949), 11.

[5] Roy J. Honeywell, "Chaplains of the Armed Force: A Compendious History," *The Military Chaplain* Volume XXX, Number 4 (April 1957), 9.

[6] Robert L. Gushwa, *The Best And Worst of Times: The United States Army Chaplaincy 1920-1945*, (Volume IV, Office of the Chief of Chaplains, Department of the Army, 1977), 113-114.

[7] Ibid.

[8] Circular Letter 232, Office of the Chief of Chaplains, August 15, 1941.

[9] Daniel B. Jorgensen, *The Service of Chaplains to Army Air Units, 1917-46*. (Wash, DC: Office, Chief of Air Force Chaplains, 1961), 75.

[10] Military Ordinariate, Circular Letter 16, (462 Madison Ave NY, NY, December 1, 1941), 1.

[11] Military Ordinariate, Circular Letter 29, 2.

[12] *The Army Builds Chapels*. Unpublished memento for the dedication of the first Army Cantonment Chapel, March 3, 1942. (Memento from Chaplain William Arnold to the Fort Myer Post Commander. Original memento with Chaplain (Colonel) Robert Nay).

[13] Ibid.

[14] Ibid.

[15] Clifford M. Drury, *The History Of The Chaplains Corps, United States Navy, Volume 2 – 1939-1949*, (U.S. Government Printing Office, Wash D.C. nd), 8.

[16] Circular Letter 249, Office of the Chief of Chaplains, April 15, 1942.

[17] Army Service Forces, Office of the Chief of Chaplains, The Chaplain Serves: Chaplain Activities, 1943. (Washington D.C. Office of the Chief of Chaplains 1 March 1944), 22-23.

[18] Circular Letter 272, Office of the Chief of Chaplains, May 1, 1943.

[19] *The Army Builds Chapels*.

Chapter 9

Policy

The one policy that is unique to the Chaplaincy within confidential communication is the privilege of non-disclosure. The Army defines policy as:

> Authoritative written guidance that affects capabilities development. When examining this DOTMLPF–P component force modernization proponents should consider any Department of Defense, interagency, or international policy issues that may prevent effective implementation of changes in the other DOTMLPF–P components.[1]

A suggested religious support policy is:

> Authoritative written guidance that affects capabilities development. This includes any Department of Defense, interagency or international policy issues that restrict or require religious support.

Policy is the implementation of civilian control over the military. From medical practice, the legal justice system, to the free exercise of religion, policy provides strict guidelines to provide for the defense of the nation while also maintaining the rule of law for the protection of the Constitution. The medical, legal, and religious support staff has certain privileges and protections. The medical and legal professions have the policy of privileged communication. However, in some circumstances, the legal and medical professions disclose communication when the circumstances or the rule of law dictate disclosure. Chaplains and religious affairs specialists[2] provide absolute privileged communication when performed as an act of religion, conscience, or as a spiritual advisor.[3] No Federal or state law or form of inquiry can force disclosure of this communication. However, this was not the case before World War II.

Confidentiality/Sacred Privileged Communication

The phrase "Tell it to the Chaplain" popularized by the World War II chaplain autobiography of the same title, captures a unique capability only a chaplain provides in the Army that helps with the harsh memories of war.[4] Whatever soldiers tell the chaplain remains between them. During this sacred privileged communication, chaplains help soldiers process and relieve their fears and disappointments by providing reminders of grace and forgiveness, and a sympathetic ear. Even non-Catholics went to Roman Catholic chaplains in the "confessional" because they knew and felt that their conversation was "protected" from investigation.[5] As the war dragged on, the Roman Catholic Military Ordinariate shared the importance of their chaplains seeking other priests for counsel and comfort by a personal account of one of their priest. "The boys have been weeping on my shoulder and now I must weep on yours."[6] The policy of absolute confidentiality was essential for every soldier, including the communication between chaplains when not in a supervisory role.

"All wars are fought twice, the first time in the battlefield, the second time in memory."[7] Absolute confidentiality between a soldier and their chaplain is a key aspect of the soldier's mental and spiritual condition, and therefore overall readiness. Since World War II, Chaplains and the health community have developed ways and means to address the harsh memories of combat. Although not the scope of this work, it is important to note that in 1974 the *Journal of Religion and Health* recognized how the "Catholic Confession" along with psychotherapy contributes to a total restoration process."[8] Also during the 1970s Protestant Chaplains, Episcopalian Chaplain (Major) Bruce Williams[9] and Southern Baptist Phil White[10] wrote about the importance of absolute confidently. The Lutheran Theologian Dietrich Bonhoeffer wrote how confessing to one another contributes to community.[11] A recent dissertation from a Roman Catholic theologian Jonathan Stotts analyzed the sacrament of Penance (confession) in theological, psychological and cultural perspectives. The author highlights a key aspect of a healthy community is a community in confession.[12]

Before World War II, the wording of Army Regulation 60-5, under Pastoral Duties, suggested that when a chaplain is directed to make "Inquiries concerning the welfare of enlisted men" the information obtained by the chaplain may be turned over to the officer directing the inquiry.[13] In all previous manuals, the term used was "confidential." For example, communication between the commander and his chaplain was confidential. Also, in all previous manuals, the chaplain may be appointed to conduct a confidential investigation, and the "chaplain has every reason to consider them as such in his intercourse with his fellow officers and soldiers." However, the command had every right to know the communication between the chaplain and the soldier.[14] This policy was in place at the start of World War II.

After two years of war, the military realized the important pastoral role chaplains performed to soldiers who experience the harsh realities of war and the value of providing confidential communication. Therefore, on March 15, 1944, the War Department issued Circular 108:

> Chaplain.- A communication from a person subject to military law, to an Army chaplain, of any denomination, made in the relationship of priest or clergyman, and penitent either as a formal act of religioun as in the confessional or one made as a matter of conscience to a chaplain in his capacity as such or as clergyman, is as a matter of policy privileged against disclosure, unless expressly waived by the individual concerned, before an investigating officer, court-martial, court of inquiry, or board of officers, or in other proceedings wherein the testimony of the chaplain is otherwise competent and admissible.[15]

The Chief of Chaplains Circular Letter, Number 284, May 1, 1944, issued the same statement.[16] Four months after the publication of the War Department Circular, *The Chaplain Manual*, published on July 5, 1944, did not include the new policy because of the rapid and changing pace of doctrine.[17] OCCH acknowledged this and quickly provided the new policy change in the circular letters, and then corrected in future Army Regulations and Chaplain Manual publications. The updated 1947 Chaplain Manual in section 63 under Pastoral Duties reflected this new order.[18] From World War II to today Army chaplains and religious affairs specialists are the only individuals who provide absolute confidentiality.

The importance of oversight of policy and its implementation is essential for any strategic leader. There were attempts to reinterpret or ridicule the policy of confidentiality. During World War II, the T. S. Card was one attempt at minimizing the importance and intent of privileged communication.

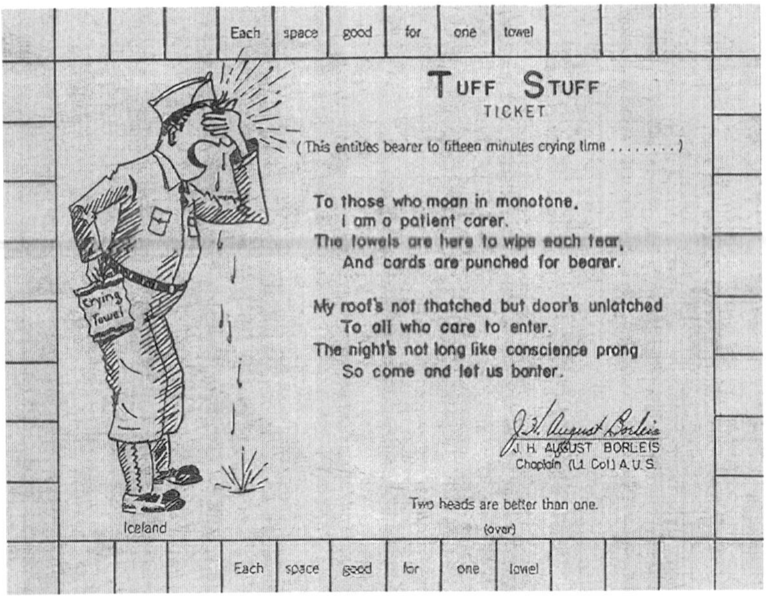

Figure 25. Card with Crying Soldier
Source: The author's personal collection.

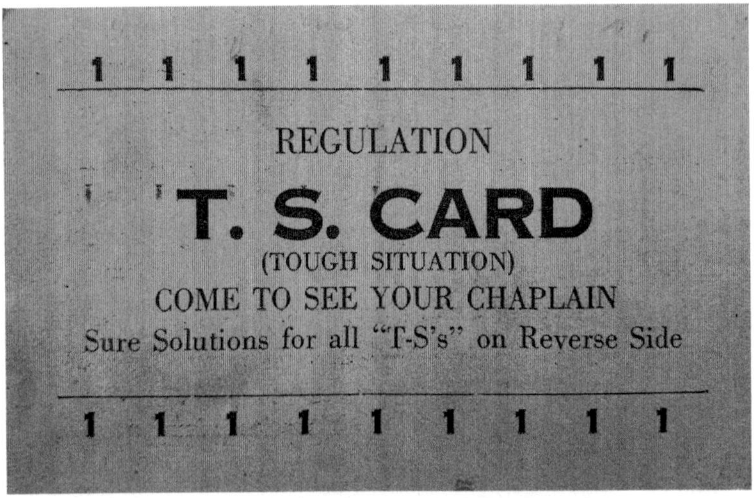

Figure 26. "Regulation" TS Card
Source: The author's personal collection.

The T. S. Card was a card developed and produced in several different variations by leaders and chaplains in the field to discourage seeing the chaplain. T. S. cards meant many things; however, the meaning was clear to the soldier. Needing to talk to the chaplain was an indication of weakness or complaining. If you needed to talk to the chaplain, you need to get over your situation and continue your mission.

One of the chaplains who did this had joined the R.O.T.C program in 1937 and then entered a seminary and graduated in 1944. He served in the 16th Armored Division, which saw minimal combat. He so boldly and with little compassion had his picture placed on the back of the T.S. for his soldiers. Chaplain Arnold corrected this chaplain and many others.

In the Addenda portion of the monthly circular letter for Commanders and Chaplains, the Chief of Chaplains provided guidance with a clear intent regarding the T.S. Card.

> It has recently come to the attention of this office that some chaplains are still resorting to the use of the so called "T.S. Card." It is believed that such a practice does not reflect favorably upon the chaplains involved or the Corps in general. It is our duty to serve with proper dignity all men who come to us irrespective of the validity of their complaints. The possibility that even a few may take offense at what would appear to be harmless humor lessens our opportunity for service and prevents us from performing as well as possible the duties for which the Army has commissioned us.[19]

Chaplains should never hinder nor look down upon a soldier who needs to speak to them, regardless of the soldier's situation. Also, showmanship and clever means that highlight the chaplain at the expense of the soldier should never be used to reach out to soldiers. Personal, private, and confidential conversations between a chaplain and a service member are precisely that, private and confidential. If soldiers suspected their chaplains shared their confidential conversations, the trust in chaplains and the care they provide for soldiers would hinder their ability to provide religious support.

A sad example of this is the Lutheran chaplain Henry Gerecke. Gerecke was one of the Army chaplains who served the war criminals at Nuremberg. Gerecke returned to the United States and sought to publish his experiences at the trials. OCCH directed him not to publish his experiences with the war criminals because of the sensitive nature of the subject and more importantly, the confidential communication between the chaplain and the defendants.

> The objection was based on the ground that the manuscript revealed intimate confidences which were deserving of the secrecy of the confessional. The War Department discourages anything that would possibly suggest to men that chaplains did not zealously guard intimate knowledge and confidence.[20]

Gerecke wrote an "unofficial report" for the Army and Navy Chaplain magazine. In this report, Gerecke revealed personal conversations and interactions with many of the defendants at Nuremburg. Gerecke disclosed that he refused Goering's request to receive Holy Communion because Goering did not profess to the doctrinal beliefs of Gerecke's denomination. Gerecke went so far to write that how Goering's young daughter wanted to meet him in heaven. After that, Goering replied, "She believes in your Saviour [*sic*], but I'll take my chances."[21] In 2014, author Tim Townsend perpetuated this embarrassment and break in confidentiality, even when Goering's daughter, Edda was still alive. Townsend took Gerecke's work, repackaged it, and sold it in 2014 as *Mission at Nuremberg: An American Army Chaplain and the Trial of the Nazis*.[22] Although the book does provide more detail about Gerecke before and after the trial, the inclusion of confidential communication between him and the defendants is a sad reminder of how one chaplain broke confidentiality.

Immediately following World War II, the policy change regarding absolute confidentiality enhanced the chaplain's competency of the nurture of the living. The Chaplain School created a new Program of Instruction (POI) titled *The Chaplain As Personal Counselor*.[23] This POI vividly explains the dynamics of the communication between the chaplain and the counselee (emphasis mine and italics retained from original).

> *Sacredness of Confidence*. The prime necessity in every interview [counseling session, ed.] is to establish permissiveness and total trust. These two elements together make up what is sometimes called *rapport*, confidence of the subject in the operator and willingness to cooperate.
>
> *Permissiveness* means that the client is completely free to speak his mind openly, unreservedly, without fear of rebuke for his frankness in telling all about himself, no matter how morally reprehensible or socially unacceptable his conduct may have been. To "accept" a situation is *not* to "approve" it; to "understand" is *not* to "condone."
>
> *The total trust* with which we are concerned here arises out of the absolute assurance which **the client must have that his "inner world" will never be discussed outside of this conference unless the**

> **client gives his permission and then only to such people as may be able to help in the case**. It is the assurance of the sacredness of confidence and the inviolability of the confidential relationship.[24]

Privileged communication replaces shame and isolation with hope and restoration.

Instead of telling soldiers "TS," chaplains at the Halloran General Hospital, Staten Island, New York went to great lengths to welcome the wounded soldiers when they returned from the battlefields of Europe. Chaplains handed out a "Welcome Card" with a calendar on one side, with the words, "Attend Religious Services Regularly. Get acquainted with your Chaplain. He is your friend." On the other side of the card was the following poem.

SEE THE CHAPLAIN

If you're puzzled and disgusted,
If you feel that you've been cheated
If you find that rules are busted-
See the Chaplain.

If you feel you are being picked on,
And your rights get circumvented
While the Law is badly dented-
See the Chaplain

Don't just stand and beef about it'
Don't just spill your grief about it;
Tell your whole belief about it-
To the Chaplain

Every new investigation
Makes for stronger congregation.
Help yourself and help the nation-
See the Chaplain![25]

Chaplain Arnold knew the power of sacred privileged communication, and when reports of the horror of war reached his office, he sought authorization to change the policy. This change empowered all chaplains to enter into this privileged role with their soldiers, which is essential to resilient soldiers and a powerful chaplain identity.

Notes

[1] U.S. Department of the Army, Army Regulation 5-22, *The Army Force Modernization Proponent System*, (Washington, DC: U.S. Department of the Army, October 28, 2015), 13.

[2] Religious Affairs Specialist is the current Army term for Chaplain Assistant.

[3] U.S. Department of the Army, Army Regulation 165-1 *Army Chaplain Corps Activities* (Headquarters, Department of the Army, Washington, D.C., June 23, 2015), 16-2.

[4] Robert Chapman, *Tell It to the Chaplain*, (Exposition Press, New York, 1952), 1.

[5] Military Ordinariate, Circular Letter 33, (462 Madison Ave NY, NY, February 27, 1943), 3.

[6] Military Ordinariate, Circular Letter 37, (June 1, 1944), 1.

[7] Viet Thanh Nguyen, *Nothing Ever Dies: Vietnam and the Memory of War*, (Harvard University Press, 2016), 4.

[8] Valerie Worthen, "Psychotherapy and Catholic Confession," *Journal of Religion and Health*, Vol 13, No. 4 (Oct., 1974) 283-284. jstor access Nov. 2, 2019.

[9] Bruce M. Williams, "Confession and Psychotherapy: Instruments of Healing," (paper submitted for Psychology and Counseling, U.S. Army Chaplain School, March 30, 1970).

[10] Phil White, "Sanctuary, Privileged Communication and the Law" (DA Pam 165-120, *Military Chaplains Review*, Winter 1979).

[11] Dietrich Bonhoeffer, *Life Together: The Classic Exploration of Christian Community* (Harper Collins, 2009), 110-122.

[12] Jonathan Andrew Stotts, "The Confessional, the Couch, and the Community: Analyzing the Sacrament of Penance in Theological, Psychological, and Cultural Perspectives" (dissertation Graduate School of Vanderbilt University, August 2016).

[13] U.S. War Department, *Chaplains*, Army Regulation 60-5, *Chaplains*, (Washington, DC: War Department, February 20, 1941), para 4.d.

[14] Technical Manual 2270-5: The Chaplain: His place and duties. 1926, page 40. Technical Manual 2270-5: The Chaplain. 1937 page 43. Technical Manual 16-205: The Chaplain. April 21, 1941, page 51.

[15] Headquarters Army Service Forces. Circular Letter 108, Office of the Chief of Chaplains, March 15, 1944.

[16] Circular Letter 284, Office of the Chief of Chaplains, May 1, 1944.

[17] U.S. War Department, *The Chaplain,* Technical Manual 16-205, (War Department, 5 July 1944), IX. 40.

[18] U.S. War Department, *The Chaplain,* Technical Manual 16-205 (War Department, June 1947), para 65.

[19] Circular Letter 293, Office of the Chief of Chaplains, February 1, 1945.

[20] Robert R. Venzke, *Confidence in Battle, Inspiration in Peace: The United States Army Chaplaincy 1945-1975 Volume V*, (Office of the Chief of Chaplains, Department of the Army, 1977), 8.

[21] Henry F. Gerecke, "Assignment With the International Tribunal As Spiritual Advisor,"

Army and Navy Chaplain Volume XVIII, Number 1, July-August 1947, p. 20.

[22] Tim Townsend, *Mission at Nuremberg: An American Army Chaplain and the Trial of the Nazis*, (William Morrow, March 11, 2014).

[23] The POI was expanded twice with greater insight on the importance and impact of absolute confidentiality. ST 16-167, "The Chaplain As Personal Counselor" (U.S. Army Chaplain School, Fort Slocum, N.Y., 1950). Department of the Army Pamphlet 16-60 The Chaplain As Counselor, (Department of the Army, April 22, 1958).

[24] *The Chaplain As Personal Counselor* (The Chaplain School, Carlisle Barracks, Pennsylvania, 1948), 5.

[25] Original card with Chaplain (Colonel) Robert Nay.

Conclusion

Thomas Carlyle's "great man theory" could apply toward Chaplain Arnold's tenure as the Army Chief of Chaplains during World War II. Carlyle proposed that great leaders are born and not made. Great leaders influence their surroundings and are the focal point of history. However, Chaplain Arnold's legacy is not the man, nor should it be. Arnold's guiding principle was sustaining every soldier's freedom to worship. His orchestration of religious support powered the successful conversion of civilian clergy to Army chaplains. He nurtured the bonds between the chaplain's initial calling and their calling to the chaplaincy. He enabled the communication between all chaplains to link them together for a common purpose, and to integrate them within the Army at every level of war. In an age of typed written documents mailed to their recipients, the monthly circular letters truly created a shared understanding of their purpose and identity as chaplains. Chaplain Arnold's legacies are the systems he created, and the people he supported. He was ahead of his time by transforming the chaplaincy across all domains of DOTMLPF-P. The three current military definitions that contributed to his success are his constant assessment of religious support operations, his talent for information and knowledge management, and his attributes as an Army leader.[1]

Chaplain Arnold continually assessed religious support operations by evaluating organizational structure, leadership ability, and resource allocation that included personnel, financial, and materiel assets. He quickly and decisively addressed any mistakes that occurred. His method of fixing problems included educating and reorienting everyone and everything toward the common purpose of providing for the free exercise of religion. Chaplain Arnold utilized Army systems and adapted his approach when necessary but never compromised his principles or his character.

To share his vision and intent for the Army chaplaincy, Chaplain Arnold mastered the tools of information and knowledge management. His ability to use procedures such as monthly reports and after action reports to collect, process, and disseminate information is impressive in an age before the internet. Chaplain Arnold used the technology of the time, but he did not become subservient to it. He used the science and technology of the printing press to create shared understanding throughout the entire chaplaincy. However, the art of what Chaplain Arnold wrote to his chaplains made the most significant impact that created a strong chaplain identity. Chaplain Arnold gave them what they needed to perform their sacred duties. Chaplains responded by providing for the free exercise of religion.

The third key ingredient that made Chaplain Arnold successful was his leadership. Chaplain Arnold, as a General Officer, embodied all the Army attributes of character, presence, intellect, leads, develops, and achieves. General Officers today receive the risk triangle as part of their training. Indicators within the triangle reveal the tendency that character, discipline, and standards do not apply to senior leaders. General Officers must adhere to the same standards of their soldiers. Chaplain Arnold did not change and compromise his principles and sacred calling, but used his rank and position to influence those around him to support the free exercise of religion.

The transition from Chaplain Arnold to his successor Chaplain Luther Miller reflected Arnold's selection in 1937 to the Chief of Chaplains. Just as Chaplain Arnold's early circular letter contained a handwritten drawing of a chaplain, in Circular Letter 296, May 1, 1945, Chaplain Arnold included for the first time a black and white picture. The picture contained Chaplain Miller, and announced him as the new Chief of Chaplains. In the next Circular Letter

297, June 1, 1945, Chaplain Miller honored his predecessor. A photo of a chaplain in a foxhole with a soldier with the words "Soldiers of God" and music staff of the *Soldiers of God* just below it. In this way, Chaplain Miller honored Chaplain Arnold by capturing his legacy.

Figure 27. Circular Letter with Soldiers of God Music Score
Source: The author's personal collection.

Chaplain Miller continued the production of the circular letters, and the letters remained a constant source for information and unity of effort. However, when Chaplain Roy Parker became Chief of Chaplains in the summer of 1949, he no longer continued the letters in the same tradition. Coincidently, the last published circular in May 1949 also occurred the same time when on May 10, 1949, the Secretary of Defense Louis Johnson issued "Transfer Order Number 35" allowing Army Chaplains to enter the new Air Force Chaplaincy.[2] These letters would have supported chaplains one year later in 1950 at the start of the Korean War. Also, the 1952 Korean War era Chaplains Field Manual contains the least amount of information for any published chaplain's manual.[3] The void this created contributed to America's Forgotten War and neglect for the service and sacrifice of many Korean War era chaplains. The monthly circular letters and other chaplain publications were an indicator of a vibrant and active Chaplaincy. In social history, there is a theory that one generation acquires the wealth, the next generation enjoys the wealth, and the third generation loses the wealth. Every generation within the chaplaincy has an opportunity to improve the legacy of the Chaplain Corps.

After World War II, the Chaplaincy's emphasis was upon the moral character of soldiers. Secretary of War Robert Patterson directed the new Chief of Chaplains, Chaplain Luther Miller, to develop a solution for the increased venereal disease rates, which were at a thirty-year high.[4] During the World War II drawdown, the chaplaincy adjusted its priorities and published the character guidance program. However, the chaplaincy neglected many of the other components of DOTMLPF-P. When there is an accomplished mission, there is a tendency to say there are no capability gaps. One domain may over perform while others are inefficient and fail to meet the standards and requirements. A leader must assess each domain independently in order to maximize all capabilities for military readiness.

Good leaders should always question assumptions knowing that enemies to national security adapt and change. Internal complacency, fear of risk, or change occurs within any bureaucracy. Bureaucrats, those who are risk averse, and minimalists look inward and are not visionary. However, there is a bias in the military toward action when there is no justif cation for change. During World War II, a justif ed requirement meant the development of a special organ, the M-45. Similarly, the Forward Thrust Doctrine dictated the need for a smaller and lighter chaplain's field kit. Does Multi-Domain Battle require the development of a new chaplain's field kit? JCIDS should provide the necessary analysis to determine future requirements. Questioning assumptions and seeking justification is essential for every domain.

Chaplain Arnold's leadership across all domains of DOTMLPF-P provided effective and efficient religious support during World War II. He truly created a sense of purpose and identity for all Army chaplains. The World War II training f lm, *For God and Country*, shows Father O'Keefe in a foxhole with several other Soldiers at night. A bomb goes off nearby. A soldier yells, "Will you cut it out for God's sake." Father O'Keefe replies, "Careful, that's my boss you are talking about." He then says, "I feel like a little prayer will help me a lot." The soldier responds, "You're not scared, are you chaplain?" Father O'Keefe says, "Yes. Afraid of missed opportunities. Afraid of all the good I could have done, neglected."[5] DOTMLPF-P is proactive instead of reactive. It is a way to assess neglected areas while providing effective and efficient opportunities to provide religious support. While Chaplain Arnold did not have the benef t of the doctrinal framework provided by DOTMLPF-P, he certainly and successfully exercised the same tenets that the doctrine would later encompass.

Notes

[1] U.S. Department of the Army, Army Regulation 600-100 *Army Profession and Leadership Policy*, (Headquarters, Department of the Army, Washington D.C. April 5, 2017), 1-11.

[2] Robert R. Venzke, *Confidence in Battle, Inspiration in Peace: The Army Chaplaincy 1945-1975*, (Volume V, Office of the Chief of Chaplains, Department of the Army, 1977), 48-49.

[3] U.S. Department of the Army, *The Chaplain*, Field Manual 16-5, (Washington, DC: Department of the Army, January 1952), 45.

[4] Venzke, 41.

[5] U.S. War Department, *For God and Country*. Training Film TF 16-2037, (U.S. War Department, Army Service Forces Signal Corps Production, 1943).

Appendix 1

Joint Capabilities Integration and Development System: Religious Support Assessment and Analysis

The Joint Capabilities Integration and Development System (JCIDS) along with the Army Force Modernization Proponent System (AFMPS) receives strategic guidance in the form of the National Security Strategy that address the global threats towards National Security. The purpose of Appendix 1 is to apply the system for religious support transformation and readiness. The system is sequential and Appendix 1 begins with the outline of the sequence. Each sequence implements strategic guidance while considering each domain of DOTMLPF-P toward the global context of threats from State and Non-State Actors, weapons of mass destruction (WMD), and the full range of military operations (ROMO).

**Army Force Modernization Proponent System:
Religious Support Assessment and Analysis**

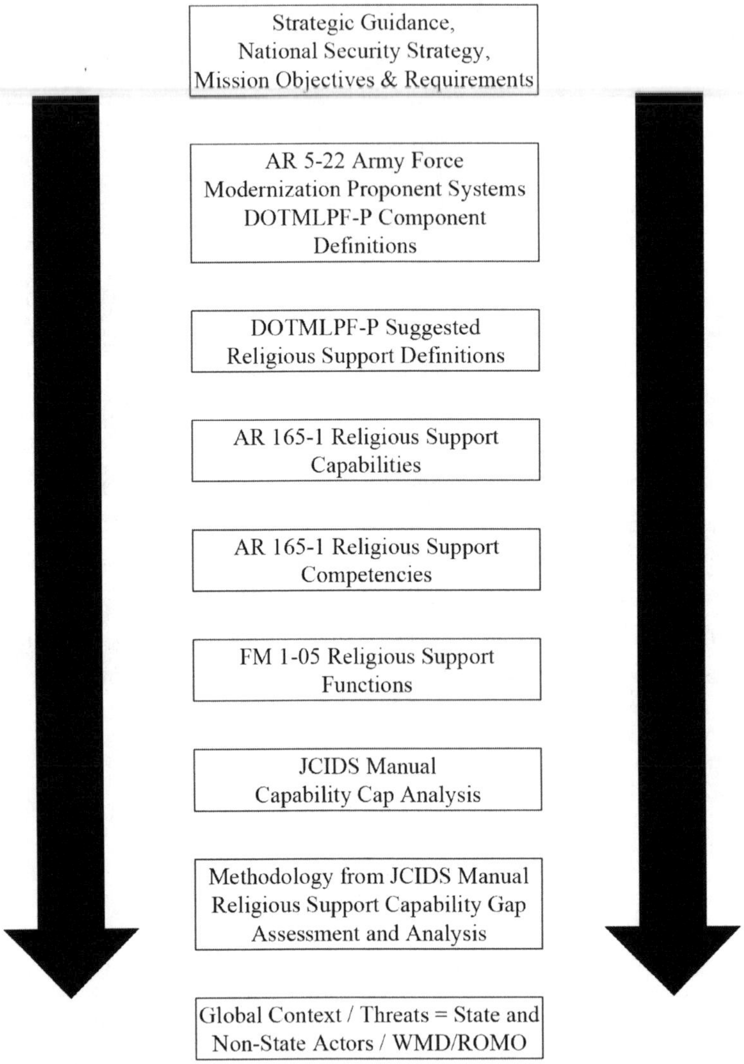

AR 5-22 Army Force
Modernization Proponent Systems
DOTMLPF-P Component Definitions

Strategic Guidance, National Security Strategy, Mission Objectives & Requirements

Doctrine: Fundamental principles by which the military forces or elements thereof guide their actions in support of national objectives.

Organization: A unit or element with varied functions enabled by a structure through which individuals cooperate systematically to accomplish a common mission and directly provide or support warfighting capabilities. Subordinate units/elements coordinate with other units/elements and, as a whole, enable the higher-level unit/element to accomplish its mission. This includes the manpower (military, civilian, and contractor support) required to operate, sustain, and reconstitute warfighting capabilities.

Training: The instruction of personnel to increase their capacity to perform specific military functions and associated individual and collective tasks.

Materiel: All items (including ships, tanks, self-propelled weapons, aircraft, and so forth, and related spares, repair parts, and support equipment but excluding real property, installations, and utilities) necessary to equip, operate, maintain, and support military activities without distinction as to its application for administrative or combat purposes.

Leader development and education: Leadership development is the product of a learning continuum that comprises training, experience, formal education, and continual self-improvement.

Personnel: The development of manpower and personnel plans, programs, and policies necessary to man, support and sustain the Army.

Facilities: Real property consisting of one or more of the following: a building, a structure, a utility system, pavement, and underlying land.

Policy: Authoritative written guidance that affects capabilities development. When examining this DOTMLPF–P component force modernization proponents should consider any Department of Defense, interagency, or international policy issues that may prevent effective implementation of changes in the other DOTMLPF–P components.

Global Context / Threats = State and Non-State Actors / WMD/ROMO

DOTMLPF-P Suggested
Religious Support Definitions

Strategic Guidance, National Security Strategy, Mission Objectives and Requirements

Global Context / Threats = State and Non-State Actors / WMD/ROMO

Doctrine: Provide Soldiers and Families with and advise Leaders on comprehensive religious support and care at every level of war in order to support national objectives and to uphold the free exercise of religion ensured by the First Amendment of the Constitution of the United States.

Organization: Chaplain Sections and Unit Ministry Teams with capabilities and competencies are integrated at every echelon, executing religious support functions, providing Unit, Area, and Denominational coverage, accountable to their assigned chain of commands and the chaplain technical staff channels to the Army Chief of Chaplains.

Training: The instruction of professional military religious leaders and professional religious advisors (56A & 56M) to increase their capacity to provide and perform with other religious and faith groups.

Materiel: All items (ecclesiastical and equipment and supplies including those in chapel facilities and in a combat environment, communication and transportation assets, administrative equipment and supplies but excluding real property such as chapels and family life centers) necessary to provide and advise religious support without distinction as to its application for administrative or combat purposes and fully integrates into all aspects of a unit's operation and function.

Leader development and education: Leader development and education strengthens and builds upon the foundational vocational calling of the chaplaincy and chaplain identity.

Personnel: Recruit, manage, and retain qualified religious support professionals IOT provide for the free exercise of religion in the Army.

Facilities: Real property such as buildings that supports the free exercise of all religious faith groups in the Army.

Policy: Authoritative written guidance that affects capabilities development. This includes any Department of Defense, interagency, or international policy issues that restrict or require religious support.

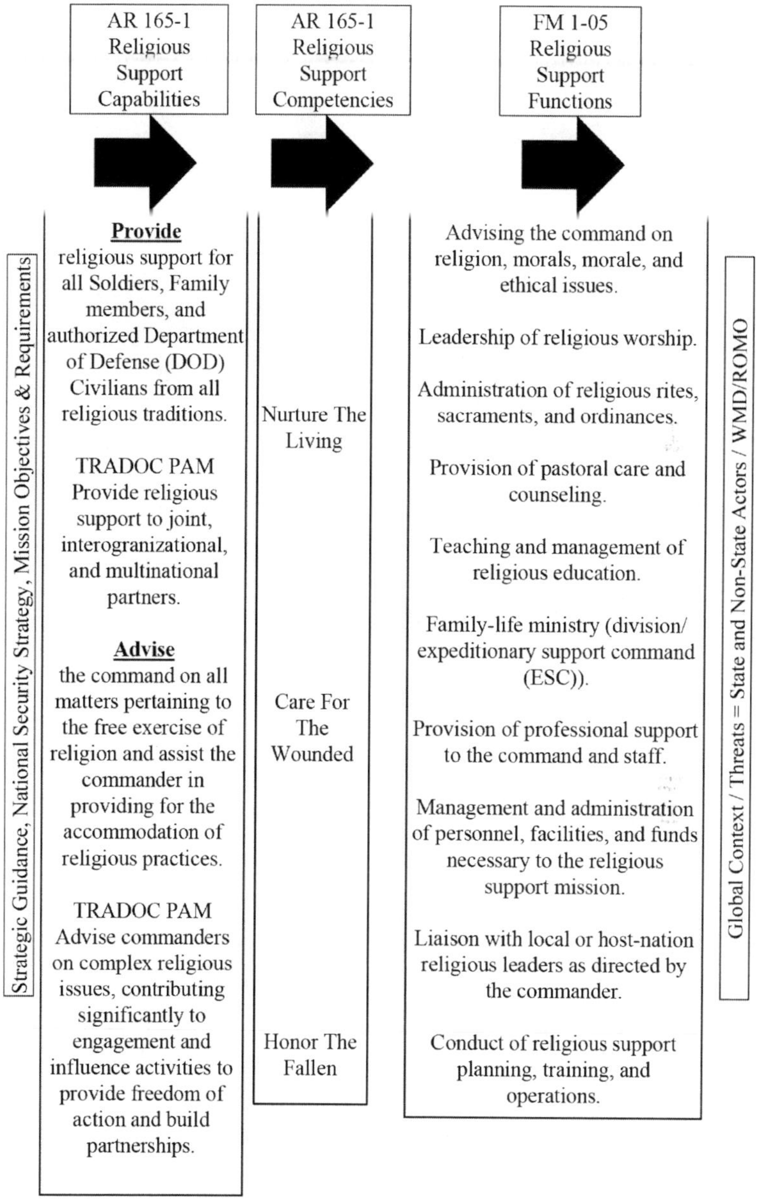
AR 165-1 Religious Support Capabilities
AR 165-1 Religious Support Competencies
FM 1-05 Religious Support Functions
Strategic Guidance, National Security Strategy, Mission Objectives & Requirements
Provide religious support for all Soldiers, Family members, and authorized Department of Defense (DOD) Civilians from all religious traditions.
TRADOC PAM Provide religious support to joint, interogranizational, and multinational partners.
Advise the command on all matters pertaining to the free exercise of religion and assist the commander in providing for the accommodation of religious practices.
TRADOC PAM Advise commanders on complex religious issues, contributing significantly to engagement and influence activities to provide freedom of action and build partnerships.
Nurture The Living
Care For The Wounded
Honor The Fallen
Advising the command on religion, morals, morale, and ethical issues.
Leadership of religious worship.
Administration of religious rites, sacraments, and ordinances.
Provision of pastoral care and counseling.
Teaching and management of religious education.
Family-life ministry (division/ expeditionary support command (ESC)).
Provision of professional support to the command and staff.
Management and administration of personnel, facilities, and funds necessary to the religious support mission.
Liaison with local or host-nation religious leaders as directed by the commander.
Conduct of religious support planning, training, and operations.
Global Context / Threats = State and Non-State Actors / WMD/ROMO

JCIDS Manual
Capability Cap Analysis

Strategic Guidance, National Security Strategy, Mission Objectives & Requirements

Lack of proficiency (inability to achieve the relevant effect in particular conditions)

Lack of sufficiency (inability to bring capable forces to bear due to force shortages or other commitments

Lack of any fielded capability solution

Need for replacement due to aging (fatigue life, technological obsolescence etc) of a fielded capability solution

Policy limitations (inability to use the force as needed due to policy constraints).

Global Context / Threats = State and Non-State Actors / WMD/ROMO

Methodology from JCIDS Manual Religious Support
Capability Gap Assessment and Analysis

Strategic Guidance, National Security Strategy, Mission Objectives and Requirements

Global Context / Threats = State and Non-State Actors / WMD/ROMO

Doctrine:
1. Does the emerging requirement or initiative enhance or limit the doctrine of providing and advising religious support for Soldiers, Families and authorized DOD Civilians?
2. What change must occur within the emerging requirement or initiative in order to provide and advise for the free exercise of religion?

Organization:
1. Are Chaplain Sections and Unit Ministry Teams with capabilities and competencies integrated at echelon with their assigned unit with a technical chain of command?
2. What change must occur within the emerging requirement or initiative in order for integrated, supervised and quality religious support that supports the free exercise of religion?

Training:
1. Are the current individual and collective training tasks sufficient at echelon?
2. What changes must occur to the individual and collective tasks lists, and what changes must occur to the annual training plan?

Materiel:
1. Does current materiel provide for the free exercise of religion for all faith groups?
2. What changes must occur within the MTOE and the supply system that allows for sufficient materiel at echelon?

Leader development and education:
1. Does the current institutional training develop competent, adaptable and agile leaders that will perform within the emerging requirement or initiative?
2. What changes must occur with institutional training and what unique challenges and requirements will Chaplains and Religious Affairs Specialists face within the Army Talent Management framework (Skill, Knowledge, Behavior)?

Personnel:
1. Is the current force structure to include policies and programs related to personnel management sufficient for the proposed emerging requirement or initiative?
2. Will changes to the TOE/TDA sustain a manpower analysis study?

Facilities:
1. Are there sufficient/functional chapels, offices and other religious support facilities that supports the emerging requirement or initiative?
2. What short term and POM solutions that will provide sacred space for the free exercise of religion?

Policy:
1. Does current DOD, DA, Chief of Chaplains policies allow for, or constrain the initiative?
2. What changes to DOD, DA, Chief of Chaplains policies must occur for sustained religious support within the emerging requirement or initiative?

Appendix 2

Design Approach to Army Chaplain History (1937-1945)

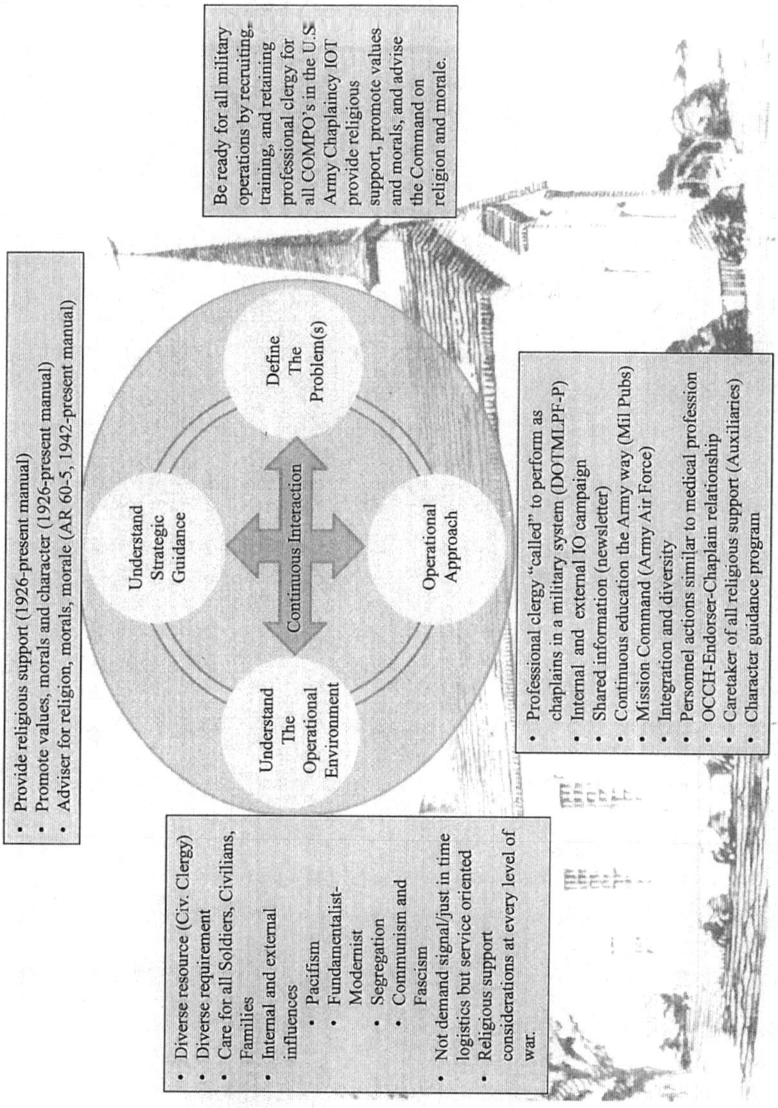

Appendix 3

How The Chaplaincy Fits Into The Army Structure Circular 266, January 15, 1943

HOW THE CHAPLAIN FITS INTO THE MILITARY PATTERN: At the present time, the Army is divided into three main sections, the Army Ground Forces, the Army Air Forces, and the Services of Supply. The first two are divided into what are known as tactical units, the last into Supply Service, Administrative Service, Service Commands, and fourteen staff agencies. The tactical units of the Army Ground Forces to which chaplains are assigned are subdivided into task forces, armies, corps, divisions, brigades, and in a few instances regiments, and battalions. The tactical units to which chaplains are attached in the Army Air Forces are task forces, air forces, commands, wings, and group. In addition to these, chaplains are attached to air base squadrons, engineer aviation battalions and regiments, signal air warning battalions, and air schools. The territorial divisions of the Services and Supply are called service commands. The continental United States is divided into nine such commands and the Military District of Washington. Under them are the reception centers (except Air Corps reception centers) where selectees report and receive their initial vaccinations and supplies, replacement training centers where men are given approximately ninety days basic training, the various services schools, including the Chaplains School, and most general hospitals. Services of Supply units are allotted chaplain on the same basis as the Army Ground Forces and the Army Air Forces.

A supervisory chaplain is assigned to each task force, army corps, and division of the Army Ground Forces, to each task force, air force, command, and wing of the Army Air Forces, and to each service command of the Services of Supply. An administrative chaplain is normally designated by the commanding officer of each post, camp, station, or school where several chaplains are assigned.

The commanding officer of each Army organization or unit is responsible for every phase of life within his command, including the religious. The chaplain is a member of his staff and is immediately responsible to him. Without trespassing upon the ecclesiastical field, the immediate commanding officer is obligated to exercise active supervision over the chaplain and his activities. From him the chaplain must seek approval of his plans for a religious program, approval to invite outside speakers, and approval to attend religious conferences and conventions on detached service. He gives official publicity to chaplain announcements and actively supports the chaplain's religious program including Divine worship.

In peacetime, chaplains are divided into three groups – Regular Army, Reserve, and National Guard. In wartime, they retain their classification but become a part of the Army of the United States. New chaplains are appointed in the Army of the United States. The work of chaplains is not determined by these classifications, and they may be transferred from unit and organization to another without regard thereto.

The work of the chaplain is very much the same as that of a pastor of a church with the exception that he is responsible for the religious care of all the men, regardless of denomination, instead of the men of a particular faith. The Protestant chaplain arranges for worship services for Catholic and Jewish men; Catholic and Jewish chaplains do the same for men not of their faiths. Where chaplains of the following major faiths are present, Catholic chaplains hold services for Catholic men, Protestant chaplains for Protestant men, and Jewish chaplains for Jewish men. Specific activities of chaplains include public worship, instruction, training, various types of daily or midweek services, pastoral duties, conferences, and visitation of men sick in the hospital or confined for infringement of regulations. As a member of the staff of the commanding officer, he also makes recommendations and reports relative to the morale and religious welfare of the command.

In addition to similar duties in behalf of headquarters units and staffs, supervisory chaplains make inspections of chaplain facilities, programs, and activities, examine monthly reports, make recommendations and reports to their immediate commanding general concerning the statue of moral and religious activities, and make recommendations for transfer or assignment of chaplains within the entire unit.

The relation of the Chief of Chaplains to the chaplain in the field is both personal and supervisory. He may correspond directly with chaplains and they with him upon subjects that relate wholly to their professional activities. He supervises their activities by inspections from his office, by the issuance of technical circulars of information and a circular letter, by the preparation of extension course material for personal study or for use in troop schools, and through the supervisory chaplain in the field. In addition to this, he is constantly alert to their needs and handicaps and is frequently able to assist in the solution of problems that apply to the entire Corps of Chaplains.

In order that the individual chaplain may more easily fit into this Army pattern, he is usually sent first to the Chaplain School where he is given basic instruction in military life, including such subjects as military correspondence, first aid, map reading, graves registration, and military law. He then reports to his assigned station and begins his work as a minister of religion in the Army.

Appendix 4

Army Chaplain Correspondence Courses

The Chaplain TM 2270-5 January 2, 1926	The Chaplain TM 2270-5 June 10, 1937	The Chaplain TM 16-205 April 21, 1941
Organization of the Army and of the Office of the Chief of Chaplains	Organization of the Army	Organization of the Army
	Office Organization and Administration, Chaplains	Office Organization and Administration, Chaplains
Administration, Discipline and Courtesies	Administration	Administration
	Military Discipline, Courtesies	Military Discipline, Courtesies
	Customs of the Service	Customs of the Service
Practical Duties of Chaplains	Practical Duties of Chaplains	Practical Duties of Chaplains
Hygiene and Sanitation	Military Sanitation and First Aid	Military Sanitation and First Aid
Military Sociology	Military Sociology	

Map Reading and Sketching	Map und Aerial Photograph Reading	Map und Aerial Photograph Reading
Military Law	Military Law-The Law of Military Offenses	Military Law-The Law of Military Offenses
	Military Law-Courts Martial	Military Law-Courts Martial
Rules of Land Warfare	Rules of Land Warfare and Military Government	Rules of Land Warfare and Military Government
	Burials and Graves Registration	Burials and Graves Registration, and Related Subjects
	Defense Against Chemical Warfare	Defense Against Chemical Warfare
	Organization of the Infantry Division	Organization of the Infantry Division (Triangular)
	Property, Emergency Procurement and Funds	
	Mobilization	
		Interior Guard Duty
		Welfare and Recreation
		Army Morale

Appendix 5

Courses of Instruction Taught at the Army Chaplain School

World War I (10 Days)	March 1942 (25 Days)	Oct 1945 (3 Months)
First aid	Military hygiene/first aid	Military sanitation
Gas defense drill	Defense against chemicals	Chemical warfare
Map reading	Map and aerial photograph reading	Map reading
Official correspondence	*Military correspondence and surveys	Army administration
Equipment	*Money and property accountability	
Customs of the service	Military discipline, customs, and courtesies	Customs and courtesies
Chaplain at the front		
Identify & burial of the dead	Graves registration	Graves Registration
Esprit de Corps	Army morale	Army morale

Lectures by auxiliaries		
French history		
French language		
Censorship		
Drill	Drill	Drill
Physical exercise	Physical exercise	Physical exercise
Devotional exercise		
Preaching	Preaching	
	*Office Organization Leadership and Administration	Army Organization
	Rules of land warfare and Field Service regulations	Military law
	**Professional duties of chaplains	**Practical duties
	*Investigations, interior guard duty, military intelligence	
	*Recreation, education, and music	Music appreciation

	*Administrative, supervisory, and cooperative duties of division, corps, and army chaplains	Administrative chaplain
		Counseling
		Staff procedures
		Special lectures
		Commandant's Time

*Chaplain Arnold directed these courses added to the Army G3 required course curriculum.

**Practical and Professional duties consisted of care for the sick, wounded, distressed, prisoners, and families.

Appendix 6

Chaplain School Student Evaluation of Teacher Form

Teaching Aptitude

Instructor: Date: Time: Class:

Elements	Rating SU	S	U	Remarks and Suggestions
1. Neatness				
2. Posture				
3. Dignity				
4. Confidence				
5. Enthusiasm				
6. Gestures				
7. Vitality				
8. Voice				
9. Speech				
10. Vocabulary				
11. Aim				
12. Motivation				
13. Organization				
14. Knowledge				
15. Method				
16. Participation				
17. Aids				
18. Blackboard				
19. Interpretation				
20. Communication				
21. Time				

22. Achievement				
23. Discipline				
24. Management				
Recommendations:				
Supervisor:				

Appendix 7

Soldiers of God: Official Chaplains' March Comparison

Original Version 1943	**Final Version 1944**
	Faithful to God, we're serving on the battlefield today. Embracing the cause of Righteousness, We're marching on our way.
Soldiers of God, we serve him faithfully, And march in His name Through thunder and flame Wherever the call may be. Trusting in God, His strength we lean upon, As into the fight the legions of light, The Soldiers of God march on.	*(Refrain)* Soldiers of God, we serve him faithfully, And march in His name Through thunder and flame Wherever the **"call"** may be. Trusting in God, His strength we lean upon, As into the fight the legions of light, The Soldiers of God march on.
	We are there, as the Chaplains of the nation, Everywhere with our fighting congregation, Serving the Lord, and serving the cause of humanity.

	Onward we go till I victory is won, For Justice and Right, the Legions of light, The Soldiers of God march on. *(Refrain)* Soldiers of God, we serve Him faithfully, And march in His name Through thunder and flame Wherever the **"call"** may be. Trusting in God, His strength we lean upon, As into the fight the legions of light, The Soldiers of God march on.
We are there, with the fighters of the nation, Everywhere at a thousand battle stations, Serving the Lord, and serving the cause of humanity. Onward we go till victory is won, For Justice and Right, the legions of light, The Soldiers of God march on.	We are there, with the fighters of the nation, Everywhere at a thousand battle stations, Serving the Lord, and serving the cause of humanity. Onward we go till victory is won, For Justice and Right, the legions of light, The Soldiers of God march on.

BIBLIOGRAPHY

Primary Sources

Army Regulations

U.S. War Department. *Army Regulation 30-1155, Quartermaster Corps: Transport Chaplain*, U.S. Government Printing Office, July 23, 1932.

———. *Army Regulation 30-1155: Quartermaster Corps: Chaplains.* U.S. Government Printing Office, July 23, 1932.

———. *Army Regulation 55-355: Transportation Corps: Chaplains.* U.S. Government Printing Office, August 22, 1942.

———. *Army Regulation 30-1805, Quartermaster: Graves Registration —General*, U.S. Government Printing Office, February 1, 1924.

———. *Army Regulation 30-1810, Quartermaster Corps: Burials outside continental limits of Unites States*, U.S. Government Printing Office, December 8, 1944.

———. *Army Regulation 30-1815, Quartermaster Corps: Reports of Burials*, U.S. Government Printing Office, February 1, 1924.

———. *Army Regulation 30-1820, Quartermaster Corps: Preparation and Disposition of Remains*, U.S. Government Printing Office, March 16, 1931.

———. *Army Regulation 30-1825, Quartermaster Corps: Disinterment of Remains*, U.S. Government Printing Office, February 1, 1924.

———. *Army Regulation 30-1830, Quartermaster Corps: Burial Expenses*, U.S. Government Printing Office, March 1, 1939.

———. *Army Regulation 30-1840, Quartermaster Corps: National Cemeteries*, U.S. Government Printing Office, January 15, 1932.

———. *Army Air Force Regulation 35-55, Personnel, Military: AAF Chaplain Program*, U.S. Government Printing Office, September 10, 1945.
———. *Army Regulation 55-355, Transportation Corps: Transport Chaplain*, U.S. Government Printing Office, August 22, 1942.
———. *Army Regulation 60-5: Chaplains: General Provisions*, U.S. Government Printing Office, February 15, 1924.
———. *Army Regulation 60-5: Chaplains: General Provisions,* U.S. Government Printing Office, February 15, 1924.
———. *Army Regulation 60-5: Chaplains: General Provisions*, U.S. Government Printing Office, December 20, 1937.
———. *Army Regulation 60-5: Chaplains: General Provisions*, U.S. Government Printing Office, February 20, 1941.
———. *Army Regulation 60-5: Chaplains: General Provisions*, U.S. Government Printing Office, November 1, 1941.
———. *Army Regulation 60-5: Chaplains: General Provisions*, U.S. Government Printing Office, May 10, 1942.
———. *Army Regulation 60-5: Chaplains: General Provisions*, U.S. Government Printing Office, December 16, 1944.
———. *Army Regulation 60-5: Chaplains: General Provisions*, U.S. Government Printing Office, December 12, 1946.
———. *Army Regulation 260-10. Flags, Colors, Standards, And Guidons*, U.S. Government Printing Office, February 8, 1923.

———. *Army Regulation 260-10 Change 3. Flags, Colors, Standards, And Guidons*, U.S. Government Printing Office, August 18, 1942.
———. *Army Regulation 260-10, Flags, Colors, Standards, Guidons, Streamers, Silver Bands, Tabards, and Automobile Place; Description and Use*, October 23, 1944.
———. *Army Regulation 350-105, Military Education: General and Special Service Schools-Designations, Locations, and Organization*, June 18, 1942. Change 1 December 31, 1942. Change 2 January 15, 1943.
———. *Army Regulation 350-150: Military Education: The Chaplains' School*. U.S. Government Printing Office, September 20, 1920.
———. *Army Regulation 350-1500: Military Education: The Chaplains' School*. U.S. Government Printing Office, August 16, 1926, 1920.
———. *Army Regulation 350-1500: Military Education: The Chaplains' School*. U.S. Government Printing Office, June 11, 1942.
———. *Army Regulation 605-10, Commissioned Officers: Officers Appointed in the Army of the United States)*, U.S. Government Printing Office, December 10, 1941.
———. *Army Regulation 615-368, Enlisted Men: Discharge, Unfitness (Undesirable habits or traits of character)*, U.S. Government Printing Office, May 14, 1947.
U.S. Department of the Army. *Army Regulation 165-1, Army Chaplain Corps Activities*, Department of the Army, Washington, D.C., June 23, 2015.

Training and Doctrine Documents

Technical Circulars

U.S. War Department. *Technical Circular No 1 Chapels*, by Office of the Chief of Chaplains. Washington, D.C., August 1, 1941.

———. *Technical Circular No 2 Division Chaplains*, by Office of the Chief of Chaplains. Washington, D.C., October 10, 1941.

———. *Technical Circular No 3 The Chaplain at the Reception Center*, by Office of the Chief of Chaplains. Washington, D.C., January 31, 1942.

———. *Technical Circular No 4 The Regimental or Unit Chaplain*, by Office of the Chief of Chaplains. Washington, D.C., March 1, 1942.

———. *Technical Circular No 5 The Chaplain in the Replacement Training Center*, by Office of the Chief of Chaplains. Washington, D.C., March 25, 1942.

———. *Technical Circular No 6 Marriage and Related Subjects*, by Office of the Chief of Chaplains. Washington, D.C., April 15, 1942.

———. *Technical Circular No 7 Chaplains' Equipment*, by Office of the Chief of Chaplains. Washington, D.C., May 1, 1942.

———. *Technical Circular No 8 The Chaplain in the Hospital*, by Office of the Chief of Chaplains. Washington, D.C., May 15, 1942.

———. *Technical Circular No 9 The Service Command Chaplain*, by Office of the Chief of Chaplains. Washington, D.C., September 19, 1942.

———. *Technical Circular No 10 The Chaplain's Assistant*, by Office of the Chief of Chaplains. Washington, D.C., November 6, 1942.

Manuals

U.S. War Department. *Technical Manual 2270-5 (draft manuscript): The* Chaplain: His place and duties, by Office of the Chief of Chaplains. Washington D.C., 1925. Transcript in the hands of Chaplain (Colonel) Robert Nay.

———. *Technical Manual 2270-5: The Chaplain: His place and duties*, by Office of the Chief of Chaplains. Washington, D.C., 1926.

———. *Technical Manual 2270-5: The Chaplain*, by Office of the Chief of Chaplains. Washington, D.C., June 10, 1937.

———. *Technical Manual 16-205: The Chaplain*, by Office of the Chief of Chaplains. Washington, D.C., April 21, 1941.

———. *Technical Manual 16-205: The Chaplain*, by Office of the Chief of Chaplains. Washington, D.C., July 5, 1944.

———. *Technical Manual 16-205: The Chaplain*, by Office of the Chief of Chaplains. Washington, D.C., June 1947.

———. *Instruction Manual For Model "G" Hammond Electronic Organs Installed in War Department Regimental Chapels*. Hammond Instrument Company, nd.

U.S. Department of the Army. *Technical Manual 10-750, Chaplains' Folding Organ*. U.S. Department of the Army, February 1948.

———. *Field Manual 16-5 The Chaplain*. United States Printing Office, January 1952.

———. *Field Manual 16-5, The Chaplain*. United States Government Printing Office, April 1958.

———. *Field Manual 16-5, The Chaplain.* United States Government Printing Office, August 1964.
———. *Field Manual 16-5, The Chaplain.* United States Government Printing Office, December 1967.
———. *Field Manual 16-5-1 (TEST): Chaplain Support TASTA-70.* United States Government Printing Office, March 1967.
———. *Field Manual 16-5, The Chaplain.* Washington, DC: Department of the Army, July 8, 1977.
———. *Field Manual 16-5, The Chaplain and Chaplain Assistant in Combat Operations.* Washington, DC: Department of the Army, December, 1984.
———. *Field Manual 16-1, Religious Support Doctrine: The Chaplain and Chaplain Assistant.* Washington, DC: Department of the Army, November, 1989.
———. *Field Manual 16-1, Religious Support.* Washington, DC: Department of the Army, May 1995.
———. *Field Manual 1-05, Religious Support.* Washington, DC: Department of the Army, April 2003.
———. *Field Manual 1-05, Religious Support.* Washington, DC: Department of the Army, October 2012.
———. *Field Manual 1-05, Religious Support.* Washington, DC: Department of the Army, January 2019.

War Department Circulars

War Department Circular No. 136, Appointment Chaplains Regular Army May 7, 1942.
War Department Circular No. 263, Appointment Chaplains Regular Army, August 5, 1942.
War Department Circular No. 23, Classification of troops, January 18, 1943. [Chaplains will not be employed as combatants, ed.]

War Department Circular No. 119, Officer Replacement pools, May 11, 1943.
War Department Circular No. 121, Appointment to Regular Army except Chaplains and Medical, May 17, 1943.
War Department Circular No. 228, Promotion of First Lieutenants, Chaplains, September 23, 1943.
War Department Circular No. 81, Chaplain Equipment, February 23, 1944.
War Department Circular No. 82, Identification Tags, February 24, 1944.
War Department Circular No. 95, Tables of Organization Chaplain, March 6, 1944.
War Department Circular No. 108, Chaplain Privileged Communication, March 15, 1944.
War Department Circular No. 204, T/O&E, May 23, 1944. [Chaplains no longer carry M3 Knife, ed.].
War Department Circular 45, Chaplain-Employment of Auxiliary Civilian Clergy, February 19, 1947.

Chief of Chaplains Circular Letters

Headquarters Army Service Forces. "Office of the Chief of Chaplains, Office Circulars 1-115", by Office of the Chief of Chaplains. Washington, D.C., November 11, 1920—November 24, 1925.
———. "Office of the Chief of Chaplains, Circular Letters 1-334", by Office of the Chief of Chaplains. Washington, D.C., January 5, 1926—May 1, 1949. [December 1, 1947 started Department of the Army "Office of the Chaplains Circular Letter 326" ed.].

Chaplain Miscellaneous Publications

Aryeh Lev ed., *What Chaplains Preach,* Command and General Staff School, 1942.

American Chaplains' of the Fifth Army. Pizzi and Pizzio, Milan, Italy, 1945.

Army Forces Pacific, *History of Chaplains' Activities in the Pacific*. Chaplain Section, Ground Headquarters, Army Forces Pacific, 1946.

Army Service Forces, Office of the Chief of Chaplains, *The Chaplain Serves: Chaplain Activities, 1943*. Washington D.C. Office of the Chief of Chaplains 1 March 1944.

Bennett, Evan L., ed., *Song and Service Book For Ship and Field*. Government Printing Office, Washington, 1942.

———. *Psalms and Hymns and Spiritual Songs: English-Chinese Edition.*

———. *Psalms and Hymns and Spiritual Songs English-Korean Edition.*

Chaplains' Association of the Army of the United States, *The Army Chaplain*. July 1, 1930-January/February 1940. [Thirty editions within 10 years. ed]

Chaplains' Association of the Army and Navy of the United States, The Army and Navy Chaplain. July/August, 1940-January/February 1948. [Thirty-two editions within 8 years. ed]

Department of the Army. *Department of the Army Pamphlet (DA PAM 16-60) The Chaplain As Counselor*. Department of the Army, April 22, 1958.

Headquarters Army Service Forces, United States Army Chaplain School, *History of USACHS: France 1918-Havard University.*

Headquarters, 7th Infantry Division. *Report Activities of Chaplains, 7th Infantry Division, Battle of Attu.* Headquarters, U.S. Army Troops, Office of the Chaplain, July 5, 1942.

Religious and Spiritual Morale Factors, The Chaplain School, Carlisle Barracks, Pennsylvania, December 1, 1948.

ST 16-155 The Chaplain and Military Morale. The Chaplain School, Carlisle Barracks, Pennsylvania, August 1950.

ST 16-152 Chaplain Supervision: General Considerations. The Chaplain School, Carlisle Barracks, Pennsylvania, June 1, 1950.

ST 16-167, "The Chaplain As Personal Counselor" (U.S. Army Chaplain School, Fort Slocum, N.Y., 1950.

The Chaplain As Personal Counselor. The Chaplain School, Carlisle Barracks, Pennsylvania, 1948.

U.S. Forces, European Theater. *Report on the Army Chaplain in the European Theater*. Study 68, Bad Nauheim, 1945.

U.S. War Department. *Helpful Hints and Other Information,* Office of the Chief of Chaplains, May 1942.

———. War Department Pamphlet 16-1, *The United States Army Chaplaincy*. Government Printing Office, August, 1946.

———. *For God and Country*. Training Film TF 16-2037, U.S. War Department, Army Service Forces Signal Corps Production, 1943.

———. War Department Pamphlet 21-39, *Conduct of a Military Funeral*. Government Printing Office, September 1947.

U.S. Department of the Army, Technical Manual 10-750 (TM) *Chaplains' Folding Organ*, Washington, DC: U.S. Department of the Army, February 1948.

Yates Julian E., and Frazier, John B. ed., *The Army and Navy Hymnal*. The Century Company, New York, 1920.

Yates Julian E., and Scott, Evan W. ed., *The Army and Navy Hymnal*. The Century Company, New York, 1925.
———. *The Army and Navy Hymnal: Hymn Edition.* The Century Company, New York, 1925.

Army Publications

U.S. War Department. *Technical Manual 10-630: Graves Registration*. U.S. Government Printing Office, September 23, 1941.
———. *Technical Bulletin 1 10-630-2: Graves Registration*. U.S. Government Printing Office, March 27, 1944.
———. *Field Manual 10-63: Graves Registration*. U.S. Government Printing Office, January 1945.
———. Technical Manual 10-285 *Deceased Personnel*, U.S. Government Printing Office, July, 1947.
———. Quartermaster Corp Manual 16-2 *Postwar Graves Registration Activities in an oversea Theater*, U.S. Government Printing Office, August, 1947.
———. Office of the Quartermaster General. *Disposition of World War II Armed Forces Dead*, U.S. Government Printing Office, 1946.
———. Mobilization Regulations 1-10 *Morale*. U.S. Government Printing Office, March 5, 1943.
———. *War Department Pamphlet No. 27-4: Procedures for Military Executions*, U.S. Government Printing Office, June 12, 1944.
———. *U.S. War Department Pamphlet No. 21-39: Conduct of a Military Funeral*, U.S. Government Printing Office, September, 1947.
U.S. Department of the Army. Army Regulation (AR) 5-22 *The Army Force Modernization Proponent System*, Department of the Army, October 28, 2015.

———. *Department of the Army Pamphlet 27-4: Procedure for Military Executions*, Department of the Army, December 1947.
Headquarters, Sixth Army. *Repatriation of World War II Dead*. UTAFP, nd.
Special Service Division. ETOUSA. *ETOUSA Special & Service Services Guide*. May 1944.
Manual for the Operation of the Joint Capabilities Integration and Development System (JCIDS), Defense Acquisition University, February 12, 2015. (accessed 14 June 2017 https://dap.dau.mil/policy/Documents/2015/JCIDS_Manual_with_errata_through_20151218.pdf).
Chairman of the Joint Chiefs of Staff Instruction 3170.001. *Joint Capabilities Integration and Development System (JCIDS)*, January 23, 2015. (accessed 14 June 2017 https://dap.dau.mil/policy/Documents/2015/CJCSI_3170_01I.pdf).

Secondary Sources

Ackermann, Henry F. *He Was Always There: The U.S. Army Chaplain Ministry In The Vietnam Conflict*. Washington, D.C.: Office of the Chief of Chaplains, 1989.
Army Chaplains: Their Hindrances and Encouragements. New York: U.S. Military Post Library Ass, 1800s.
Arnold, William R. "The Army Chaplain," Army and Navy Journal, LXXX, 81, LXXXI, 76, LXXXII, 104.
———. "Faith under Fire; What Kind of Boy will the Army Send Back?" *Homan's Home Companion*, LXXI, 24.
———. "Our Fighting Troops Receive Spiritual Encouragement from Army Chaplains," *Victory Bulletin*, V, 7.

———. "Wanted: Army Chaplains" *Christian Advocate*, CX\ III, 1233.
———. "We Are Strong in Spirit," *Country Gentlemen*, CXII, 10.
Axton, John T. Jr. *Brief History of Chaplains in U.S. Army*, Fort Leavenworth, Kansas: General Service Schools, 1925.
Barish, Louis, ed. *Rabbis in Uniform: A Century of Service to God and Country 1862-1962*. Jonathan David: New York, 1962.
Brasted. Alva J. *Suggestions for newly appointed chaplains*, np, 1918.
———. *Service to Service Men*, Augsburg Publishing House, 1941.
Brinsfield, John W. *Encouraging Faith; Supporting Soldiers: The United States Army Chaplaincy, 1975-1995*. Washington, DC: Office of Chief of Chaplains, 1997.
Brinsfield, John W., Gaylord Gunhus, Kenneth Lawson, and Douglas McCullough, *Courageous in Spirit; Compassionate in Service: The Gunhus Years (1999-2003)*. Washington, DC: Office of the Chief of Chaplains, 2004.
Brinsfield, John W. and Kenneth Lawson, *A History of the Army Chaplaincy: The Hicks Years 2003-2007.* Washington, DC: Office of the Chief of Chaplains, 2007.
Brown, Henry A. *Army and Navy Church Service Manual: Adapted for Use in the United States Army and Navy.* New York: U.S. Soldiers Christian Aid Association, 1901.
Brown, W. Y. *The Army Chaplain: His Office, Duties, and Responsibilities, and the Means of Aiding Him.* Philadelphia: William S. & Alfred Martien, 1863.
Budd, Richard M. *Serving Two Masters: The Development of American Military Chaplaincy, 1860-1920*. Lincoln, Nebr: University of Nebraska Press, 2002.

Chapman, Robert. *Tell it to the Chaplain*. Exposition Press, New York, 1952.

Clemens, Joseph. *Duties and Privileges of Chaplains*, Manuscript submitted to the War D, 1914.

Converse, Elliott Vanveltner. *Rearming for the Cold War, 1945-1960, History of acquisition in the Department of Defense, v. 1*. Washington, D.C. Historical Office, Office of the Secretary of Defense, 2012.

Crosby, Donald F. SJ. *Battlefield Chaplains: Catholic Priests in World War II*. University Press of Kansas, 1994.

Cross, Christopher with William R. Arnold. *Soldiers of God*. New York: Concordia Publishing House, 1945.

Drury, Clifford M., *The History Of The Chaplains Corps, United States Navy, Volume 2 – 1939-1949*, U.S. Government Printing Office, Wash D.C., 1949.

Faust, Drew Gilpin. *This Republic of Suffering: Death and the American Civil War*. Knopf, January 8, 2008.

Gellerman, Robert F. *The American Reed Organ and the Harmonium*. Vestal Press, January 1, 1997.

Gushwa, Robert L. *The Best and Worst of Times: The United States Army Chaplaincy, 1920-1945, Vol. IV*. Washington, DC: Office of the Chief of Chaplains, 1977.

Hammond, Jonathan Pinkney. *The Army Chaplain's Manual, Designed As a Help to Chaplains in the Discharge of Their Various Duties*. Philadelphia: J.B. Lippincott, 1863.

Haney, M.L. *The Story of my Life: An Autobiography*. Published by the author, Normal, Illinois, 1904.

Headley, Joel Tyler. *The Chaplains and Clergy of the Revolution*. New York: C. Scribner, 1864.

Hirrel, Leo P. Army Sustainment "The Beginnings of the Quartermaster Graves Registration Service" PB 700-14-04, Volume 46, Issue 4, July-August 2014.

Honeywell, Roy J. *Chaplains of the United States Army*. Wash, DC: Office, Chief of Chaplains, 1958.

Jorgensen, Daniel B. *The Service of Chaplains to Army Air Units, 191746, Volume I*. Wash, DC: Office, Chief of AF Chaplains, 1961.

Jorgensen, Daniel B. *Air Force Chaplains, 1947-1960, Volume II*. Wash, DC: Office, Chief of AF Chaplains, 1961.

Loesser, Frank. *Praise the Lord and Pass The Ammunition!!*. Famous Music Corp, 1942.

May, Earl Chapin. *Century of Silver, 1847-1947, Connecticut Yankees and a Noble Metal*. Robert M. McBride & Company, New York, 1947.

Nance, Ellwood C., ed. *Faith of our Fathers*. Bethany Press, 1944.

Ossad, Steven L. and Marsh, Don R. *Major General Maurice Rose: World War II's Greatest Forgotten Commander*. Taylor Trade Publishing, 2003.

McCracken, Elizabeth. *Triptychs for Servicemen*, The Living Church (Morehouse-Gorham Co., January 30, 1944.

Military Ordinariate, Circular Letters 1-45, (462 Madison Ave NY, NY).

Moss, James A. *Manual of Military Training*. George Banta Publishing Company, Army and College Printers, Menasha Wisconsin, 1914.

Moss, James A. *Manual of Military Training (Second Revised Edition)*. George Banta Publishing Company, Army and College Printers, Menasha Wisconsin, 1917.
Nance, Ellwood C., ed. *Faith of our Fathers*. Bethany Press, 1944.
Nave, Orville J. *Nave's Handbook on the Army Chaplaincy: With a Supplement on the Duty of the Churches to Aid the Chaplains by Follow-Up Work in Conserving the*

Moral and Religious Welfare of the Men Under the Colors. Los Angeles, Calif: [s.n.], 1917.

Nay, Robert. "The Operational, Social, and Religious Influence Upon the Army Chaplain Field Manual, 1926-1952." Master's thesis, Command and General Staff College, 2008.

Norton, Herman Albert. *Struggling for Recognition: The United States Army Chaplaincy, 1791-1865. Vol II*. Washington: Office of the Chief of Chaplains, Dept. of the Army, 1977.

Poling, Daniel A. *My Eyes Have Seen: An Autobiography*. McGraw-Hill, 1959.

Reston, James B. *Prelude To Victory*. Alfred A. Knopf New York, 1942.

Reichley, James. *Religion in American Public Life*. Washington D.C: Brookings Institution, 1985.

Simpson, G. W. *Manual for U.S. Army Chaplains*. [Washington, DC?]: [s.n.], 1893.

Slomovitz, Albert Isaac. *The Fighting Rabbis: Jewish Military Chaplains and American History*. New York University Press, 1999.

Snape, Michael. *God and Uncle Sam: Religion and America's Armed Forces in World War II*. Boydell Press, May 21, 2015.

Stahl, Ronit Y. *Enlisting Faith: How the Military Chaplaincy Shaped Religion and State in Modern America*. Harvard University Press, November 6, 2017.

Steward, T. G. and G. W. Prioleau. *Active Service, or, Religious Work Among U.S. Soldiers. A Series of Papers by Our Post and Regimental Chaplains*. New York: United States Army Aid Association, 1890s.

Stover, Earl F. *Up From Handymen: The United States Army Chaplaincy, Vol III*. Washington, D.C.: Office of the Chief of Chaplains, Department of the Army, 1977.

Thompson, Donald A. *American Army Chaplaincy: A Brief History*. Wash, DC: Chaplains Assn, 1946.

Thompson, Parker C. *The United States Army Chaplaincy. Vol. 1, From its European Antecedents to 1791.* Washington DC: Office of the Chief of Chaplains, 1978.

Venzke, Rodger R. *Confidence in Battle, Inspiration in Peace, The United States Army Chaplaincy 1945-1975, Vol V*. Washington, D.C.: Office of the Chief of Chaplains, 1977.

Walters, Kevin L. *Beyond The Battle: Religion and American Troops in World War II*. University Of Kentucky, 2013.

Waring, George J. *Chaplain's Duties and How Best to Accomplish His Work*. Washington [D.C.]: G.P.O., 1912.

———. *Regulations for Catholic Chaplains in the United States Army and Navy*. S.l: s.n, 1918.

———. *United States Catholic Chaplains in the World War*. New York city: Ordinariate, Army and Navy Chaplains, 1924.